Lecture Notes in Computer Science　　8585

Commenced Publication in 1973
Founding and Former Series Editors:
Gerhard Goos, Juris Hartmanis, and Jan van Leeuwen

More information about this series at http://www.springer.com/series/7409

Tilmann Rabl · Nambiar Raghunath
Meikel Poess · Milind Bhandarkar
Hans-Arno Jacobsen · Chaitanya Baru (Eds.)

Advancing
Big Data Benchmarks

Proceedings of the 2013 Workshop Series
on Big Data Benchmarking
WBDB.cn, Xi'an, China, July 16–17, 2013
and WBDB.us, San José, CA, USA
October 9–10, 2013
Revised Selected Papers

 Springer

Editors

Tilmann Rabl
Hans-Arno Jacobsen
University of Toronto
Toronto, ON
Canada

Nambiar Raghunath
Cisco Systems, Inc.
San José
USA

Meikel Poess
Oracle Corporation
Redwood Shores
USA

Milind Bhandarkar
Pivotal Software, Inc.
Palo Alto
USA

Chaitanya Baru
University of California at San Diego
La Jolla
USA

ISSN 0302-9743
ISBN 978-3-319-10595-6
DOI 10.1007/978-3-319-10596-3

ISSN 1611-3349 (electronic)
ISBN 978-3-319-10596-3 (eBook)

Library of Congress Control Number: 2014952779

LNCS Sublibrary: SL3 – Information Systems and Applications, incl. Internet/Web, and HCI

Springer Cham Heidelberg New York Dordrecht London

Printed on acid-free paper

Springer is part of Springer Science+Business Media (www.springer.com)

Preface

Formed in 2012, the Big Data Benchmarking Community (BDBC) represents a major step in facilitating the development of benchmarks for objective comparisons of hardware and software systems dealing with emerging Big Data applications. Led by Chaitanya Baru, Tilmann Rabl, Meikel Poess, Milind Bhandarkar, and Nambiar Raghunath, the BDBC has successfully conducted four international Big Data Benchmarking Workshops (WBDB) bringing industry experts and researchers together to present and debate challenges, ideas, and methodologies to benchmark big data systems. The first WBDB was held in San José, California during May 8–9, 2012 at the Brocade Executive Briefing Center and hosted by Brocade. The second WBDB was held in Pune, India during December 17–18, 2012, hosted by Persistent Systems and Infosys.

This book contains the joint proceedings of the Third and Fourth Workshops on Big Data Benchmarking. The third WBDB was held in Xi'an, China, during July 16–17, 2013, hosted by Xi'an University of Posts and Telecommunications and Mellanox. The fourth WBDB was again held in San José, California from October 9–10, 2013 and hosted by Brocade.

The 2013 editions of the workshop series saw a much clearer focus on the benchmarking topics and also a much higher number of concrete benchmark proposals. In this book, we have collected seven proposals for big data benchmarks, which are collected in Section "Big Data Benchmarks". Yet there are still new and exciting application descriptions that are presented in Section "Application and Scenarios". Additionally, in Section "Tools, Systems, and Surveys", we have collected an overview of different systems and applications and strategies for an industry standardization approach within the TPC.

The 15 papers in this book were selected out of a total of 33 presentations in WBDB2013.cn and WBDB2013.us. All papers were reviewed in two rounds. Shepherds helped the authors to incorporate the feedback from reviewers as well as the audience at the workshop. We thank the sponsors, members of the program committee, authors, and participants for their contributions to these workshops. The hard work and close cooperation of a number of people have been critical to the success of the WBDB workshop series.

May 2014

Tilmann Rabl
Nambiar Raghunath
Meikel Poess
Milind Bhandarkar
Hans-Arno Jacobsen
Chaitanya Baru

WBDB 2012 Organization

General Chairs

Chaitanya Baru   San Diego Supercomputer Center, USA
Tilmann Rabl   University of Toronto, Canada

Program Committee

Milind Bhandarkar   Pivotal, USA
Dhruba Borthakur   Facebook, USA
Yanpei Chen   Cloudera, USA
John Galloway   SalesForce, USA
Ahmad Ghazal   Oracle, USA
Boris Glavic   IIT, USA
Bhaskar Gowda   Intel, USA
Eyal Gutkind   Mellanox Technologies, USA
Songlin Hu   Chinese Academy of Science, China
Hans-Arno Jacobsen   MSRG, Canada
Jian Li   IBM, USA
Tong Liu   Mellanox Technologies, USA
Nambiar Raghunath   Cisco, USA
D.K. Panda   Ohio State University, USA
Meikel Poess   Oracle, USA
Francois Raab   InfoSizing, USA
Kai Sachs   SAP, Germany
Florian Stegmaier   University of Passau, Germany
Suresh Vobbilisetty   Brocade, USA
Jerry Zhao   Google, USA
Jianfeng Zhan   Chinese Academy of Sciences, China
Xiaohui Zhou   Xi'an University of Posts
                                    and Telecommunications, China

WBDB 2012 Sponsors

WBDB2013.cn Sponsors

NSF
Mellanox
Brocade
Pivotal
NetApp

WBDB2013.us Sponsors

Brocade
Seagate
Pivotal
NetApp

Contents

Tools, Systems, and Surveys

Big Data Benchmarks

A BigBench Implementation in the Hadoop Ecosystem

Badrul Chowdhury[1], Tilmann Rabl[1]([✉]), Pooya Saadatpanah[2], Jiang Du[2], and Hans-Arno Jacobsen[1]

[1] Middleware Systems Research Group, University of Toronto, Toronto, Canada
badrul.chowdhury@mail.utoronto.ca, tilmann.rabl@utoronto.ca,
jacobsen@eecg.toronto.edu
http://msrg.org
[2] Database Research Group, University of Toronto, Toronto, Canada
{pooya,jdu}@cs.toronto.edu

Abstract. BigBench is the first proposal for an end to end big data analytics benchmark. It features a rich query set with complex, realistic queries. BigBench was developed based on the decision support benchmark TPC-DS. The first proof-of-concept implementation was built for the Teradata Aster parallel database system and the queries were formulated in the proprietary SQL-MR query language. To test other systems, the queries have to be translated.

In this paper, an alternative implementation of BigBench for the Hadoop ecosystem is presented. All 30 queries of BigBench were realized using Apache Hive, Apache Hadoop, Apache Mahout, and NLTK. We will present the different design choices we took and show a proof of concept evaluation.

1 Introduction

Big data analytics is an ever growing field of research and business. Due to the drastic decrease of cost of storage and computation more and more data sources become profitable for data mining. A perfect example is online stores, while earlier online shopping systems would only record successful transactions, modern systems record every single interaction of a user with the website. The former allowed for simple basket analysis techniques, while current level of detail in monitoring makes detailed user modeling possible.

The growing demands on data management systems and the new forms of analysis have led to the development of a new breed of systems, big data management systems (BDMS) [1]. Similar to the advent of database management systems, there is a vastly growing ecosystem of diverse approaches. This leads to a dilemma for customers of BDMSs, since there are no realistic and proven measures to compare different offerings. To this end, we have developed Big-Bench, the first proposal for an end to end big data analytics benchmark [2]. BigBench was designed to cover essential functional and business aspects of big data use cases.

© Springer International Publishing Switzerland 2014
T. Rabl et al. (Eds.): WBDB 2013, LNCS 8585, pp. 3–18, 2014.
DOI: 10.1007/978-3-319-10596-3_1

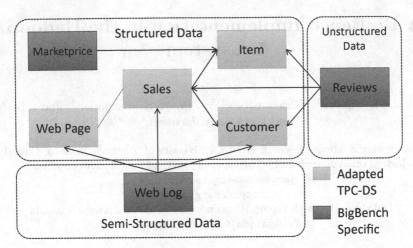

Fig. 1. BigBench schema

In this paper, we present an alternative implementation of the BigBench workload for the Hadoop eco-system. We re-implemented all 30 queries and ran proof of concept experiments on a 1 GB BigBench installation.

The rest of the paper is organized as follows. In Sect. 2, we present an overview of the BigBench benchmark. Section 3 introduces the parts of the Hadoop ecosystem that were used in our implementation. We give details on the transformation and implementation of the workload in Sect. 4. We present a proof of concept evaluation of our implementation in Sect. 5. Section 6 gives an overview of related work. We conclude with future work in Sect. 7.

2   BigBench Overview

BigBench is an end-to-end big data analytics benchmark, it was built to resemble modern analytic use cases in retail business. As basis for the benchmark, the Transaction Processing Performance Council's (TPC) new decision support benchmark TPC-DS was chosen [3]. This choice highly sped up the development of BigBench and made it possible to start from a solid and proven foundation. A high-level overview of the data model can be seen in Fig. 1. The TPC-DS data model is a snowflake schema with 6 fact tables, representing 3 sales channels, store sales, catalog sales, and online sales, each with a sales and a returns fact table. For BigBench the catalog sales were removed, since they have decreasing significance in retail business. As can be seen in Fig. 1, additional big data specific dimensions were added. Marketprice is a traditional relational table storing competitors prices. The Web Log portion represents a click-stream that is used to analyze the user behavior. This part of the data set is semi-structured, since different entries in the weblog represent different user actions and thus have different format. The log is generated in form of an Apache Web server log.

The unstructured part of the schema is generated in form of product reviews. These are, for example, used for sentiment analysis. The full schema is described in [4].

BigBench features 30 complex queries, 10 of which are taken from TPC-DS, the others were specifically developed for BigBench. The queries are covering major areas of big data analytics as specified in [5]. As a result, the queries cannot be expressed by pure SQL queries since they include machine learning techniques, sentiment analysis, and procedural computations. In Teradata Aster, this is solved using built in functions that are internally processed in a MapReduce fashion. The benchmark, however, does not dictate a specific implementation, therefore, the benchmark can be implemented in various ways. The full list of queries can be found in [4].

3   Technologies for BigBench on Hadoop

In this section, the technologies used to create an open-source implementation of BigBench are described. BigBench is mainly implemented using four open-source software frameworks: Apache Hadoop, Apache Hive, Apache Mahout, and the Natural Language Processing Toolkit (NLTK). We used the following versions for our implementation: Apache Hadoop V0.20.2, Apache Hive 0.8.1, Apache Mahout 0.6, and NLTK 3.0.

3.1   Hadoop

Apache Hadoop provides a scalable distributed file system and features to perform analysis on and store large data sets using the MapReduce framework [6]. Its architecture consists of many components (see Fig. 2) and a discussion of the design decisions with implementation details can be found [7]. Only components that are most relevant to the BigBench implementation will be described in the following.

The *Hadoop Distributed File System* (HDFS) is modeled after the Unix file system hierarchy with 3-way replication of data for security and analysis performance purposes. The Hadoop command line interface provides access to a most

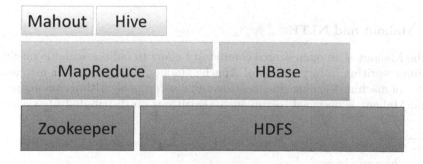

Fig. 2. Hadoop stack

standard Unix file operations such as *ls, rm, cp*, etc. A complete reference can be found on Apache Hadoop's website[1].

A cluster implementing HDFS has 3 main components: *HDFS client, namenode*, and *datanode*. The namenode primarily stores meta data. It keeps a record of the *namespace tree*, which stores information relevant to file block allocation to datanode. It should be noted that all of the namespace data is stored in RAM. There can only be one namenode in any single cluster in the version of Hadoop used. However, on the other hand, there are usually multiple datanode in a cluster. Each datanode contains two files in the local file system: one to store metadata and the other to store the actual data. The HDFS client provides an interface for user-created applications to access and modify HDFS. Access is provided in a two-tiered process: first, the metadata in the namenode is extracted and then information is used to access the relevant datanodes.

3.2 Hive

Apache Hive is a data warehousing solution being developed with a view to provide traditional SQL programmers access to the functionality of MapReduce [8]. To this end, Hive features an SQL-like structured language called Hive Query Language (HiveQL)[2]. The version used in the BigBench implementation supports a proper subset of standard SQL operations in addition to providing the facility to "stream" custom user-defined MapReduce jobs to perform operations on stored tables.

We will only outline the basic architecture here, a more detailed description can be found in [8]. Some components of the Hive architecture that are worth of noticing are the *metastore*, the *query compiler*, and the *execution engine*. The metastore is used to store all operation-critical information pertaining to table schema, table location, table columns and their data types. The information stored in the metastore needs to be extracted very quickly for data analysis and transformation and, therefore, it is usually stored in a local database. The most commonly used metastore system is MySQL. The query compiler compiles the queries written in Hive-QL and optimizes the queries where possible. Finally, the execution engine that is based on MapReduce executes the tasks specified by the compiled queries according to their precedence in the dependency tree.

3.3 Mahout and NLTK

Apache Mahout is an open-source community effort to build a scalable machine learning algorithm library on top of Apache Hadoop[3]. It has an ever-increasing number of machine learning classification and clustering algorithms among many others. Mahout is designed to run such algorithms on distributed file systems.

---

[1] http://hadoop.apache.org/
[2] http://hive.apache.org/
[3] https://mahout.apache.org/

In the BigBench implementation, Mahout is used mainly to run the *k-means* algorithm on HDFS.

NLTK provides a library of Python functions for the processing of natural language using standard statistical techniques [9]. It is distributed under the Apache License[4]. NLTK is used to implement sentiment analysis in BigBench queries.

4 Query Implementation

In this section, the implementation of the different flavors of queries will be discussed. The overall number of 30 queries has been grouped into 4 categories: Pure Hive queries, Hive queries with MapReduce programs, Hive queries using natural language processing, and queries using Apache Mahout. In the following, we will give an example for each of the different flavors of queries.

The distribution of the different query types is shown in Table 1. Table 2 shows the data types that the queries access as specified in Sect. 2.

It should be noted that queries that use NLTK and Mahout also require preprocessing by Hive. Therefore, Apache Hive is critical to all data processing activities in this implementation of BigBench.

4.1 Loading Data

The synthetically generated data is loaded onto Hive in 2 steps:

1. Each table required by the benchmark is created in Hive using the syntax shown in Listing 1.1. The full list of table descriptions is presented in [4].
2. The data is loaded onto the Hive tables using the directive in Listing 1.2.

Data is loaded in the same way for all of the 24 tables of the benchmark.

Table 1. Distribution of BigBench queries by query type

Data type	BigBench queries	Percentage
Pure Hive	5, 6, 7, 9, 11, 12, 13, 14, 16, 17, 21, 22, 23, 24	46.7
Hive + MR	1, 2, 3, 4, 8	16.7
Hive + Hadoop Streaming	15, 18	6.7
Mahout	20, 25, 26, 29, 30	16.7
NLTK	10, 19, 27, 28	13.2

Table 2. Distribution of BigBench queries by data type

Data type	BigBench queries	Percentage
Structured	1, 6, 7, 9, 13, 14, 15, 16, 17, 19, 20, 21, 22, 23, 24, 25, 26, 29	60.0
Semi-structured	2, 3, 4, 5, 8, 12, 30	23.3
Unstructured	10, 11, 18, 27, 28	16.7

---

[4] http://www.nltk.org/

```
drop table if exists customer_demographics;

create table customer_demographics
( cd_demo_sk              bigint,
  cd_gender               string,
  cd_marital_status       string,
  cd_education_status     string,
  cd_purchase_estimate    int,
  cd_credit_rating        string,
  cd_dep_count            int,
  cd_dep_employed_count   int,
  cd_dep_college_count    int
)
ROW FORMAT DELIMITED FIELDS TERMINATED BY '|';
```

Listing 1.1. Creating the *customer_demographics* table in Hive

```
LOAD DATA LOCAL INPATH 'customer_demographics.dat'
    OVERWRITE INTO TABLE customer_demographics;
```

Listing 1.2. Loading *customer_demographics* data onto Hive

4.2   Hive Queries

Hive does not support the full standard SQL syntax; some of these peculiarities will be discussed in more detail in this section as relevant to BigBench Query 21. *Query 21 (TPC-DS 29)*:

> Get all items that were sold in stores in a given month and year and which were returned in the next six months and re-purchased by the returning customer afterwards through the web sales channel in the following three years. For those these items, compute the total quantity sold through the store, the quantity returned and the quantity purchased through the web. Group this information by item and store.

```
SELECT i_item_id, i_item_desc, s_store_id, s_store_name,
       sum(ss_quantity) AS store_sales_quantity,
       sum(sr_return_quantity) AS store_returns_quantity,
       sum(ws_quantity) AS web_sales_quantity
  FROM store_sales, store_returns, web_sales, date_dim d1,
       date_dim d2, date_dim d3, store, item
 WHERE d1.d_moy             = 4
   AND d1.d_year            = 1998
   AND d1.d_date_sk         = ss_sold_date_sk
   AND i_item_sk            = ss_item_sk
   AND s_store_sk           = ss_store_sk
   AND ss_customer_sk       = sr_customer_sk
   AND ss_item_sk           = sr_item_sk
   AND ss_ticket_number     = sr_ticket_number
   AND sr_returned_date_sk  = d2.d_date_sk
   AND d2.d_moy             BETWEEN 4 AND  4 + 3
   AND d2.d_year            = 1998
   AND sr_customer_sk       = ws_bill_customer_sk
   AND sr_item_sk           = ws_item_sk
   AND ws_sold_date_sk      = d3.d_date_sk
   AND d3.d_year            IN (1998,1998+1,1998+2)
 GROUP BY i_item_id, i_item_desc, s_store_id, s_store_name
 ORDER BY i_item_id, i_item_desc, s_store_id, s_store_name;
```

Listing 1.3. Query 21 - Aster SQL-MR Syntax

```
SELECT * FROM
  SELECT i.i_item_id AS item_id, i.i_item_desc AS item_desc,
         s.s_store_id AS store_id, s.s_store_name AS store_name,
         SUM(ss.ss_quantity) AS store_sales_quantity,
         SUM(sr.sr_return_quantity) AS store_returns_quantity,
         SUM(ws.ws_quantity) AS web_sales_quantity
    FROM store_sales ss
         JOIN item i ON (i.i_item_sk = ss.ss_item_sk)
         JOIN store s ON (s.s_store_sk = ss.ss_store_sk)
         JOIN date_dim d1 ON (d1.d_date_sk = ss.ss_sold_date_sk
              AND d1.d_moy = 4 AND d1.d_year = 1998)
         JOIN store_returns s ON (ss.ss_customer_sk = sr.sr_customer_sk
              AND ss.ss_item_sk = s.sr_item_sk)
         JOIN date_dim d2 ON (sr.sr_returned_date_sk = d2.d_date_sk
              AND d2.d_moy > 4-1 AND d2.d_moy < 4+3+1 AND d2.d_year = 1998)
         JOIN web_sales ws ON (sr.sr_item_sk = ws.ws_item_sk)
         JOIN date_dim d3 ON (ws.ws_sold_date_sk = d3.d_date_sk
              AND d3.d_year IN (1998 ,1998+1 ,1998+2))
  GROUP BY i.i_item_id, i.i_item_desc, s.s_store_id, s.s_store_name)
  ORDER BY item_id, item_desc, store_id, store_name;              .
```

Listing 1.4. Query 21 - Hive Syntax

Query 21 is taken from TPC-DS and thus a traditional relational query. The version in SQL-MR syntax is shown in Listing 1.3. It joins 8 tables, 3 fact tables and 5 dimensions. The largest table, store_sales is joined with item, store, and date_dim to find items bought in a particular month. Then, by joining with store_returns, the items that were returned are filtered. Finally, using a join with web_sales items that were previously returned in store and then bought again online within 3 years are selected. These are grouped by item and store. The Hive version, as shown in Listing 1.4, is an almost word-for-word translation of the SQL-MR version with a few notable differences in syntax. The Hive implementation makes use of the JOIN syntax extensively; this is due to the fact that the current Hive version only supports a single table in the FROM-clause, which may be a composite of multiple tables itself, as in this case. Also, arguments of the WHERE-clause in the SQL-MR query have been used as JOIN conditions in the Hive version on grounds of improving efficiency: the Hive version's implementation eliminates the need to loop over the resulting table again after the joins have taken place.

4.3   Hive and MapReduce

Data processing tasks becomes simpler and more intuitive using custom programs in several BigBench queries. External programs are therefore used in the queries by means of Hive's program streaming feature to mimic some of SQL-MR's built-in functions. This is discussed with on the example of Query 10 of BigBench. Query 10 is an example of sentiment analysis, which is not included in the standard SQL functionality.

Query 10:

> For all products, extract sentences from its product reviews that contain
> positive or negative sentiment and display the sentiment polarity of the
> extracted sentences.

```
SELECT pr_item_sk, out_content, out_polarity, out_sentiment_words
  FROM ExtractSentiment
       (ON product_reviews100
        TEXT_COLUMN ('pr_review_content')
        MODEL ('dictionary')
        LEVEL ('sentence')
        ACCUMULATE ('pr_item_sk')
        )
 WHERE out_polarity = 'NEG'
    OR out_polarity = 'POS';
```

Listing 1.5. Query 10 - SQL-MR Syntax

```
ADD FILE mapper_10.py;
ADD FILE reducer_10.py;

FROM (
  FROM    product_reviews
    MAP     product_reviews.pr_item,
            product_reviews.pr_review_content
    USING 'python mapper_10.py'
    AS item, polarity
) mapper

REDUCE mapper.item, mapper.polarity
  USING 'python reducer_10.py'
  AS (item STRING, polarity STRING);
```

Listing 1.6. Query 10 - Hive Syntax

In the mapper and reducer files are imported using the ADD FILE filename
directive. The mapper is invoked using the USING LANGUAGE MAPPER_FILE directive; the reducer is used similarly. The mapper should generate a (KEY,
VALUE) vector as per map-reduce guidelines: this vector is generated using the
MAP KEY, VALUE directive. Later, the REDUCE KEY, VALUE directive is used to
reduce the mapped vector.

4.4   Hive and Natural Language Processing

All of the natural language processing capabilities of the Hive implementation
of BigBench are implemented using the Natural Language Toolkit (NLTK) 3.0
package. In this section, the NLP-processing capability of Query 10 is analyzed
in more detail.

All of the features of the NLTK package are available in the program after
it is imported using the standard Python import syntax.

```
import nltk
[..]
def get_word_features(wordlist):
  wordlist = nltk.FreqDist(wordlist)
  word_features = wordlist.keys()
  return word_features

def extract_sentiment(tweet):
  [..]
  for (words, sentiment) in pos_tweets + neg_tweets:
      words_filtered = [e.lower() for e in words.split() if len(e) >= 3]
      tweets.append((words_filtered, sentiment))
  training_set = nltk.classify.apply_features(extract_features, tweets)
  classifier = nltk.NaiveBayesClassifier.train(training_set)
  return classifier.classify(extract_features(tweet.split()))
[..]
```

Listing 1.7. Query 10 - Sentiment_Analysis.py Program

The package provides powerful APIs for natural language processing. An excellent example of such an API is `nltk.FreqDist(DOCUMENT)`, as used in the sentiment analysis in Listing 1.7. The `FreqDist` method gets a Python set as parameter and returns the modified set containing the frequency distribution of each word in the original input set. The training and application of the NLTK classifier on a body of text is done in several steps. First, the `nltk.classify.apply_features()` method is used to apply a "negative" or "positive" label to each feature of the training data (the `extract_features` method extracts the features from the list of tweets). Then, the classifier is trained using Naive Bayes by invoking the method `nltk.NaiveBayesClassifier train(.training_set)`. Finally, the trained classifier is used to label the features of any new body of text by invoking `classifier.classify(extract_features (tweet.split()))`.

The overall quality of the results depends on the size and quality of the training data. In the current version of the implementation, the model is trained on a very small hand-made data set; a future improvement to the model will train it on larger training sets to improve its repertoire of feature labels.

Programs to analyze natural language have thereby been used (using streaming in Hive) to add features such as sentiment analysis to some Hive queries in this implementation.

4.5 Mahout

Apache Mahout is used to implement all of the machine-learning capabilities of BigBench; this will be exemplified with reference to Query 20 of BigBench in this section. The SQL-MR implementation of Query 20 can be seen in Listing 1.8. *Query 20*:

Customer segmentation for return analysis: Customers are separated along the following dimensions: return frequency, return order ratio (total number of orders partially or fully returned versus the total number

of orders), return item ratio (total number of items returned versus
the number of items purchased), return amount ration (total monetary
amount of items returned versus the amount purchased), return order
ratio. Consider the store returns during a given year for the computation.

```
CREATE VIEW sales_returns AS (
  SELECT s.ss_sold_date_sk AS s_date,
         r.sr_returned_date_sk AS r_date,
         s.ss_item_sk AS item,
         s.ss_ticket_number AS oid,
         s.ss_net_paid AS s_amount,
         r.sr_return_amt AS r_amount,
         (CASE WHEN s.ss_customer_sk IS NULL
               THEN r.sr_customer_sk ELSE s.ss_customer_sk END) AS cid,
         s.ss_customer_sk AS s_cid,
         sr_customer_sk AS r_cid
    FROM store_sales s LEFT JOIN store_returns100 r ON
            s.ss_item_sk = r.sr_item_sk
         AND s.ss_ticket_number = r.sr_ticket_number
   WHERE s.ss_sold_date_sk IS NOT NULL);

CREATE VIEW clusters AS (
  SELECT cid,
         100.0 * COUNT (DISTINCT (CASE WHEN r_date IS NOT NULL
                                       THEN oid ELSE NULL END))
             / COUNT (DISTINCT oid) AS r_order_ratio,
         SUM (CASE WHEN r_date IS NOT NULL THEN 1 ELSE 0 END)
             / COUNT (item) * 100 AS r_item_ratio,
         SUM (CASE WHEN r_date IS NOT NULL THEN r_amount ELSE 0 END)
             / SUM (s_amount) * 100 AS r_amount_ratio,
         COUNT (DISTINCT (CASE WHEN r_date IS NOT NULL
                               THEN r_date ELSE NULL END))
                               AS r_freq
    FROM sales_returns
   WHERE cid IS NOT NULL
   GROUP BY 1
  HAVING COUNT (DISTINCT (CASE WHEN r_date IS NOT NULL
                               THEN r_date ELSE NULL END)) > 1);

SELECT *
  FROM kmeans (ON
        (SELECT 1)
        PARTITION BY 1
        DATABASE ('benchmark')
        USERID ('benchmark')
        PASSWORD ('benchmark')
        INPUTTABLE ('clusters␣AS␣c')
        OUTPUTTABLE ('user_return_groups')
        NUMBERK('4'));

SELECT clusterid, cid
  FROM kmeansplot (ON
        clusters AS c
        PARTITION BY ANY
        ON user_return_groups dimension
        CENTROIDSTABLE ('user_return_groups'))
  ORDER BY clusterid, cid;

DROP TABLE user_return_groups;
DROP VIEW clusters;
DROP VIEW sales_returns;
```

Listing 1.8. Query 20 - SQL-MR Syntax

```
CREATE VIEW IF NOT EXISTS sales_returns AS
  SELECT s.ss_sold_date_sk AS s_date,
    r.sr_returned_date_sk AS r_date,
    s.ss_item_sk AS item,
    s.ss_ticket_number AS oid,
    s.ss_net_paid AS s_amount,
    r.sr_return_amt AS r_amount,
    (CASE WHEN s.ss_customer_sk IS NULL
          THEN r.sr_customer_sk ELSE s.ss_customer_sk END) AS cid,
    s.ss_customer_sk AS s_cid,
    sr_customer_sk AS r_cid
  FROM store_sales s
  LEFT OUTER JOIN store_returns r ON s.ss_item_sk = r.sr_item_sk AND
    s.ss_ticket_number = r.sr_ticket_number
  WHERE s.ss_sold_date_sk IS NOT NULL;

CREATE TABLE IF NOT EXISTS all_sales_returns AS
  SELECT * FROM sales_returns;

CREATE VIEW IF NOT EXISTS clusters AS
  SELECT cid,
    100.0 * COUNT(distinct(CASE WHEN r_date IS NOT NULL
                                THEN oid ELSE NULL end))
         / COUNT(distinct oid) AS r_order_ratio,
    SUM(CASE WHEN r_date IS NOT NULL
             THEN 1 ELSE 0 END)
         / COUNT(item)*100 AS r_item_ratio,
    SUM(CASE WHEN r_date IS NOT NULL
             THEN r_amount ELSE 0.0 END)
         / SUM(s_amount)*100 AS r_amount_ratio,
    COUNT(distinct (CASE WHEN r_date IS NOT NULL
                         THEN r_date ELSE NULL END)) AS r_freq
  FROM all_sales_returns
  WHERE cid IS NOT NULL
GROUP BY cid;

DROP TABLE IF EXISTS twenty;

CREATE TABLE IF NOT EXISTS twenty
  ROW FORMAT DELIMITED
  FIELDS TERMINATED BY''
  LINES TERMINATED BY'\n'
  STORED AS TEXTFILE LOCATION'/mahout_io/twenty/' AS
  SELECT * FROM clusters;

DROP TABLE IF EXISTS all_sales_returns;
```

Listing 1.9. Query 20 - Part 1 in Hive Syntax

The Hive/Mahout implementation of Query 20 can be seen in Listings 1.9 and 1.10. First the table **clusters** is created using Hive. It contains the fields on which segmentation analysis will be performed, namely **return_order_ratio**, **return_item_ratio**, **return_amount_ratio**, **return_frequency**. It is stored as a single whitespace character-delimited text file which is used as the input file to the Mahout k-means program that is shown in Listing 1.10.

```
public class Twenty {
  public static void writePointsToFile( List<Vector> points, String fileName,
      FileSystem fs, Configuration conf ) throws IOException {
    Path path = new Path(fileName);
    SequenceFile.Writer writer = new SequenceFile.Writer(fs, conf, path,
      LongWritable.class, VectorWritable.class);
    long recNum = 0;
    VectorWritable vec = new VectorWritable();
    for ( Vector point : points ) {
```

```
    vec.set(point);
    writer.append(new LongWritable(recNum++), vec);
  }
  writer.close();
}

public static List<Vector> getPoints( double[][] tuples ) {
  List<Vector> points = new ArrayList<Vector>();
  for ( int i = 0; i < tuples.length; i++ ) {
    double[] fr = tuples[i];
    Vector vec = new RandomAccessSparseVector(fr.length);
    vec.assign(fr);
    points.add(vec);
  }
  return points;
}

public static void main( String args[] ) throws IOException {
  [..]
  // number of centres is 4 as per Query 20
  int k = 4;
  List<Vector> vectors = getPoints(myPoints);
  File tuples = new File("tuples");
  if ( !tuples.exists() ) {
    tuples.mkdir();
  }
  tuples = new File("tuples/points");
  if ( !tuples.exists() ) {
    tuples.mkdir();
  }
  Configuration conf = new Configuration();
  FileSystem fs = FileSystem.get(conf);
  writePointsToFile(vectors,"tuples/points/file1", fs, conf);
  Path path = new Path("tuples/clusters/part-00000");
  SequenceFile.Writer writer = new SequenceFile.Writer(fs, conf,
    path, Text.class, Cluster.class);
  for ( int i = 0; i < k; i++ ) {
    Vector vec = vectors.get(i);
    Cluster cluster = new Cluster(vec, i, new EuclideanDistanceMeasure());
    writer.append(new Text(cluster.getIdentifier()), cluster);
  }
  writer.close();
  KMeansDriver.run(conf, new Path("tuples/points"), new
    Path("tuples/clusters"), new Path("output"), new
    EuclideanDistanceMeasure(), 0.001, 10, true, false);
  SequenceFile.Reader reader = new SequenceFile.Reader(fs,
    new Path("output/" + Cluster.CLUSTERED_POINTS_DIR + "/part-m-00000"),
    conf);
  IntWritable key = new IntWritable();
  WeightedPropertyVectorWritable value = new
      WeightedPropertyVectorWritable();
  [..]
  reader.close();
  }
}
```

Listing 1.10. Query 20 - Part 2 as a Mahout Program

The k-means clustering algorithm in Mahout has a very specific control flow. First, each input tuple (which is a single line in the input *clusters* file) is converted into a vector; the cardinality of the tuple is preserved by this operation. In the Java program, the getPoints() method does this by creating a RandomAccessSparseVector for each tuple. Then, these vectors are written to a file in the specified input directory; the writePointsToFile() method does this creating VectorWritable and SequenceFileWriter objects to create the writable representation of the vector and to perform the write-operation respectively. In this process, a Hadoop-specific data type –LongWritable– is used. Finally, the points are clustered in several passes over the input vectors and the output after each pass is stored in separate subdirectories within the output directory. In the program, this final step is commenced by running the static function run() from the KMeansDriver class, which takes in the similarity measure to be used to perform the clustering (in this case, the EuclideanDistanceMeasure) as a parameter.

The program is called from the command line using Hadoop streaming; the program is run as shown in Listing 1.11. However, before using the Hadoop streaming feature, the respective Mahout libraries must be added to the HADOOP_CLASSPATH environment variable. The exact libraries to be included for this particular program are shown in Listing 1.11.

```
$ export
    HADOOP_CLASSPATH=~/mahout/mahout-core-0.6.jar:~/mahout/mahout-math-0.
    6.jar:~/mahout/mahout-core-0.6-job.jar
$ /home/usr/hadoop-0.20.2/bin/hadoop jar Twenty.jar Twenty
    /mahout_io/Twenty/000000_0 /mahout_io/Twenty
```

Listing 1.11. Running K-means Program of Query 20 Using Hadoop Streaming

5   Evaluation

We run a proof-of-concept evaluation on a single setup. The system was fitted with 5.8 GB of RAM, a 750 GB SATA hard-disk, and a 3.2 GHz Intel Xeon Quadcore processor. Each query was run on a 1.3 GB data set, the results are shown in the below table. The *System Runtime* attribute corresponds to the time obtained by using the Unix *time* command when running the queries. The *Reported Time* attribute corresponds to the time reported by Hive and Mahout respectively; it should be noted that for queries which require running multiple engines, the numbers in the table correspond to the sum of the partial running time of each engine.

The results are plotted in Fig. 3. In general, the Unix *time* utility reported a higher time than the internal time reporting feature of Hive and Mahout. The internal time reporting feature of both Hive and Mahout displays the time it takes to complete a specified Map-Reduce job, so the difference between the two depicted times corresponds to the *set-up* and *tear-down* time of the system for each query.

In particular, Query 13 takes a noticeably longer time than the other queries due to the presence of many JOIN statements. Each JOIN statement is translated to a map-reduce job by Hive based on the join condition, and since the query was run on a single node, the query execution engine incurred considerable loss in time due to the *set-up* and *tear-down* time consumed by each individual job. Also, the tables that are joined in this query are much richer in content than in the other queries.

Query	System Runtime/s	Reported Time/s	Query	Actual Runtime/s	Reported Time/s
1	97	96	2	58	56
3	38	37	4	279	275
5	34	33	6	294	264
7	298	282	8	51	47
9	185	174	10	13	11
11	69	60	12	105	174
13	4694	4373	14	219	216
15	844	783	16	170	109
17	349	347	18	642	548
19	91	88	20	206	157
21	273	271	22	293	292
23	634	498	24	195	193
25	202	160	26	62	60
27	55	52	28	32	30
29	294	288	30	301	296

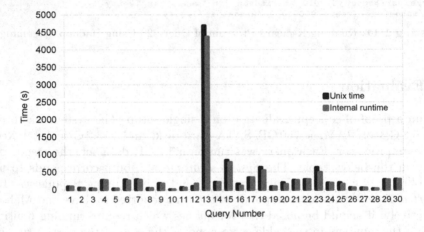

Fig. 3. Query Runtime

6   Related Work

TPC benchmarks are commonly used for benchmarking big data systems. For big data analytics, TPC-H and TPC-DS are obvious choices and TPC-H has been implemented in Hadoop, Pig[5], and Hive[6] [10,11]. A subset of TPC-DS has recently been used to compare DBMSes with Impala and Hive[7]. However, TPC-H and TPC-DS are pure SQL benchmarks and thus do not cover all the different aspects that MapReduce systems are typically used for. Several proposals try to modify TPC-DS similar to BigBench to cover typical big data use

---

[5] https://issues.apache.org/jira/browse/PIG-2397
[6] https://issues.apache.org/jira/browse/HIVE-600
[7] http://blog.cloudera.com/blog/2014/01/impala-performance-dbms-class-speed/

cases. Zhao et al. propose Big DS, which extends the TPC-DS model with social marketing and advertisement [12]. Currently, Big DS is in a very early design stage and no query set and data model are available. Once the benchmark has matured, it should be possible to complement BigBench with the Big DS proposal. Another TPC-DS modification is proposed by Yi and Dai as part of the HiBench suite [13,14]. The authors use the TPC-DS model to generate web logs similar to BigBench. Unlike BigBench the authors use this for an ETL process. This again is orthogonal to BigBench and can be included in future work. There have been several other proposals, most of which are component benchmarks testing specific functions of the big data systems. Two notable examples are the Berkeley Big Data Benchmark[8] and the benchmark presented by Pavlo et al. [15]. Another example is BigDataBench, which is a suite similar to HiBench and mainly targeted at hardware benchmarking [16]. Although interesting and very useful, these benchmarks do not present an end to end scenario and thus have another focus than BigBench.

7   Conclusion

BigBench is the only fully specified end to end benchmark for big data analytics currently available. In this paper, we presented details about our ongoing implementation for the Hadoop ecosystem. The implementation is completely based on open-source libraries and frameworks typically used in big data deployments. The queries and the data set can be downloaded from the MSRG website[9].

For future work, we will work on improving the data generation and the data model. We are currently building a complete kit for measuring the end to end processing time including loading and refresh. We will investigate the inclusion of other proposals such as the ETL-pipeline proposed as part of the HiBench suite [14].

References

1. Carey, M.J.: BDMS performance evaluation: practices, pitfalls, and possibilities. In: Nambiar, R., Poess, M. (eds.) TPCTC 2012. LNCS, vol. 7755, pp. 108–123. Springer, Heidelberg (2013)
2. Ghazal, A., Rabl, T., Hu, M., Raab, F., Poess, M., Crolotte, A., Jacobsen., H.A.: BigBench: towards an industry standard benchmark for big data analytics. In: Proceedings of the ACM SIGMOD Conference (2013)
3. Pöss, M., Nambiar, R.O., Walrath, D.: Why you should run TPC-DS: a workload analysis. In: VLDB, pp. 1138–1149 (2007)
4. Rabl, T., Ghazal, A., Hu, M., Crolotte, A., Raab, F., Poess, M., Jacobsen, H.-A.: BigBench specification V0.1. In: Rabl, T., Poess, M., Baru, C., Jacobsen, H.-A. (eds.) WBDB 2012. LNCS, vol. 8163, pp. 164–201. Springer, Heidelberg (2014)

---

[8] https://amplab.cs.berkeley.edu/benchmark/
[9] http://msrg.org

5. Manyika, J., Chui, M., Brown, B., Bughin, J., Dobbs, R., Roxburgh, C., Byers, A.H.: Big data: the next frontier for innovation, competition, and productivity. Technical report, McKinsey Global Institute (2011). http://www.mckinsey.com/ insights/mgi/research/technology_and_innovation/big_data_the_next_frontier_for_ innovation

6. Dean, J., Ghemawat, S.: MapReduce: simplified data processing on large clusters. Commun. ACM **51**(1), 107–113 (2008)

7. Shvachko, K., Kuang, H., Radia, S., Chansler, R.: The Hadoop distributed file system. In: 26th IEEE Symposium on Mass Storage Systems and Technologies, pp. 1–10 (2010)

8. Thusoo, A., Sarma, J.S., Jain, N., Shao, Z., Chakka, P., Anthony, S., Liu, H., Wyckoff, P., Murthy, R.: Hive: a warehousing solution over a Map-Reduce framework. Proc. VLDB Endow. **2**(2), 1626–1629 (2009)

9. Bird, S., Klein, E., Loper, E., Baldridge, J.: Multidisciplinary instruction with the natural language toolkit. In: Proceedings of the Third Workshop on Issues in Teaching Computational Linguistics, TeachCL '08, pp. 62–70 (2008)

10. Moussa, R.: TPC-H benchmark analytics scenarios and performances on Hadoop data clouds. In: Benlamri, R. (ed.) NDT 2012, Part I. CCIS, vol. 293, pp. 220–234. Springer, Heidelberg (2012)

11. Kim, K., Jeon, K., Han, H., Kim, S., Jung, H., Yeom, H.: MRBench: a benchmark for MapReduce framework. In: 14th IEEE International Conference on Parallel and Distributed Systems, 2008, ICPADS '08, December 2008, pp. 11–18 (2008)

12. Zhao, J.M., Wang, W., Liu, X.: Big data benchmark - Big DS. In: Rabl, T., Raghunath, N., Meikel, P., Milind, B., Jacobsen, H.-A., Chaitanya, B. (eds.) WBDB 2013. LNCS, vol. 8585, pp. 49–57. Springer, Heidelberg (2014)

13. Huang, S., Huang, J., Dai, J., Xie, T., Huang, B.: The HiBench benchmark suite: characterization of the MapReduce-based data analysis. In: ICDEW (2010)

14. Yi, L., Dai, J.: Experience from hadoop benchmarking with HiBench: from micro-benchmarks toward end-to-end pipelines. In: Rabl, T., Raghunath, N., Meikel, P., Milind, B., Jacobsen, H.-A., Chaitanya, B. (eds.) WBDB 2013. LNCS, vol. 8585, pp. 43–48. Springer, Heidelberg (2014)

15. Pavlo, A., Paulson, E., Rasin, A., Abadi, D.J., DeWitt, D.J., Madden, S., Stonebraker, M.: A comparison of approaches to large-scale data analysis. In: SIGMOD '09: Proceedings of the 35th SIGMOD International Conference on Management of Data, pp. 165–178 (2009)

16. Wang, L., Zhan, J., Luo, C., Zhu, Y., Yang, Q., He, Y., Gao, W., Jia, Z., Shi, Y., Zhang, S., Zhen, C., Lu, G., Zhan, K., Li, X., Qiu, B.: BigDataBench: a big data benchmark suite from internet services. In: Proceedings of the 20th IEEE International Symposium on High Performance Computer Architecture. HPCA (2014)

A Mid-Flight Synopsis of the BG Social Networking Benchmark

Shahram Ghandeharizadeh$^{(\boxtimes)}$ and Sumita Barahmand

Database Laboratory, Computer Science Department,
USC, Los Angeles, CA 90089-0781, USA
{shahram,barahman}@usc.edu

Abstract. BG is a benchmark that rates the performance of a data store for processing interactive social networking actions such as view a member's profile, invite a member to be friends, accept a friend request, and others. It is motivated by a proliferation of data stores from a variety of academic and industrial contributors including social networking companies, e.g., Voldemort by LinkedIn. BG is designed to provide a system architect with insights into alternative design principles such as the use of a weak consistency technique instead of a strong one, different physical data models such as relational and JSON, factors that impact vertical and horizontal scalability of a data store, the consistency versus availability tradeoff in the CAP theorem, among others. While BG is a recently introduced benchmark (less than a year old as of this writing), it combines elements of maturer benchmarks and extends them to simplify its use by the practitioners and experimentalists. This paper provides a synopsis of the BG benchmark by identifying its strengths and limitations in our daily use cases. The identified limitations shape our research activities and the obtained solutions shall be incorporated into future BG releases. Thus, this workshop paper is a mid-flight glimpse into our current research efforts with BG.

1   Introduction

In an article that appeared in the July 2012 issue of the Communications of the ACM, David Patterson observes when a discipline has good benchmarks, debates are settled and the discipline makes rapid progress [16]. Today, we have an abundance of architectures for data stores and services with only a handful of benchmarks to substantiate their many claims. Academia, cloud service providers such as Google and Amazon, social networking sites such as LinkedIn and Facebook, and computer industry continue to contribute systems and services with novel architectures and assumptions. In 2010, Rick Cattell surveyed 23 systems [7] and we are aware of 10 new[1] ones since that writing. In his survey, Cattell identified a "gaping hole" with a scarcity of benchmarks to substantiate

---

[1] Apache Jackrabbit and RavenDB, Titan, Oracle NoSQL, FoundationDB, STSdb, EJDB, FatDB, SAP HANA, CouchBase.

© Springer International Publishing Switzerland 2014
T. Rabl et al. (Eds.): WBDB 2013, LNCS 8585, pp. 19–31, 2014.
DOI: 10.1007/978-3-319-10596-3_2

the claims made by the different systems. We have designed and implemented a social networking benchmark named BG [2] (visit http://bgbenchmark.org) to address certain aspects of the hole that is too large to address with just one benchmark.

BG's workload consists of actions that either read or write a small amount of data from big data, typically termed simple operations [10,21]. Today's BG is designed for high throughput data stores that provide interactive response times. A long term objective is to extend BG with complex analytics that require processing of a large amount of data using machine learning algorithms. This would make BG suitable to evaluate Hadoop and other implementation of the MapReduce [8] framework, see Sect. 7 for details.

We developed BG in 2012 and released a stable version of it in January 2013. Its conceptual schema and eleven actions are an abstraction of today's social networking sites such as Google+, Facebook and others. In [2], we provide a comprehensive list of the surveyed sites and a matrix that describes compatibility of BG's actions with those supported by a site. Figure 1 shows BG's conceptual schema. The concept of members with registered profiles befriending one another is at the core of this schema. Its implementation in a physical data store is dictated by an experimentalist. BG is data store agnostic and one may tailor both the physical schema and the implementation of actions to highlight the strengths of a data store. At the time of this writing, an implementation of BG's schema and actions is available for the following data stores:

– MySQL, Oracle, PostgreSQL, and VoltDB as relational data stores.
– MongoDB and CouchBase as document stores.
– Microsoft Azure as a cloud service provider.
– Hibernate as an Object Relational Mapping (ORM) framework.
– Cache augmented data stores with memcached, EhCache, Twemcache, and KOSAR.

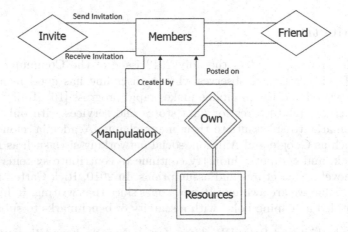

Fig. 1. Conceptual design of BG's database.

Table 1. Three mixes of social networking actions.

BG social actions	Type	Very low	Low	High
		(0.1%) Write	(1%) Write	(10%) Write
View Profile (VP)	Read	40%	40%	35%
List Friends (LF)	Read	5%	5%	5%
View Friends Requests (VFR)	Read	5%	5%	5%
Invite Friend (IF)	Write	0.04%	0.4%	4%
Accept Friend Request (AFR)	Write	0.02%	0.2%	2%
Reject Friend Request (RFR)	Write	0.02%	0.2%	2%
Thaw Friendship (TF)	Write	0.02%	0.2%	2%
View Top-K Resources (VTR)	Read	40%	40%	35%
View Comments on Resource (VCR)	Read	9.9%	9%	10%
Post Comment on a Resource (PCR)	Write	0%	0%	0%
Delete Comment from a Resource (DCR)	Write	0%	0%	0%

The first column of Table 1 shows the eleven actions that constitute BG. The name of the actions is self explanatory. All actions that reference members are binary consuming two member ids as input. For example, the two member ids specified with the View Profile action identify the member who is viewing a profile and the member whose profile is being viewed. Those actions that consume a resource id either read the resource and its comments or modify a comment on that unique resource. The different actions may either read or write data from a data store as highlighted by the second column of Table 1. We refer the interested reader to [2] for a detailed description of each action.

The last three columns of Table 1 show three different workloads corresponding to a very low, low, and a high percentage of write actions. (According to Facebook, more than 99% of its workload consists of read actions [9].) Each column specifies a fixed percentage of occurrence for each action in a workload. We use the presented three workloads in our experiments on a daily basis. All three are symmetric workloads that cause an experiment to complete with approximately the same number of confirmed friendships and pending friendships as those in the beginning of the experiment. The number of these relationships is impacted by the frequency of the following actions: Invite Friend, Accept Friend Request, Reject Friend Request, and Thaw Friendship actions. For this number to remain unchanged, the rate at which BG generates friendships should equal the rate at which it thaws friendships. This is realized by satisfying the following two conditions: (1) percentage of Thaw Friendship and Accept Friend Request must be identical, and (2) percentage of Invite Friend must equal the sum of percentage of Reject Friend Request and Accept Friend Request. In Table 1, the frequency of Post Comment on a Resource (PCR) and Delete Comment from a Resource (DCR) are intentionally kept at zero to demonstrate that one may specify workloads consisting of either a single or a few actions. When specifying frequencies of PCR and DCR, a symmetric workload should define the same frequency of occurrence for each action.

BG is a stateful benchmark that generates valid actions. For example, it extends a friendship from Member A to Member B only when they are not friends. It realizes this by maintaining a representation of the social graph in its memory. BG uses this representation to ensure its emulated simultaneous members and resources are unique at an instance in time.

A novel feature of BG is its ability to quantify the amount of unpredictable (stale, inconsistent, erroneous) data produced by a data store. BG evaluates a data store for a workload that specifies an SLA. An example SLA may require 95 % of actions to be performed faster than 100 milliseconds with no more than 0.01 % unpredictable data for $\Delta = 10$ min. BG includes a heuristic search technique to quantify the maximum throughput (actions per second) observed with a data store while satisfying the pre-specified SLA. This is termed the Social Action Rating, SoAR, of the data store.

We have employed BG to investigate design and implementation of novel architectures for data intensive applications, e.g., to compare a relational representation of a social graph with its JSON representation [5], quantify the trade-offs associated with alternative consistency techniques for a cache augmented relational data store [12], and others. In these use cases, we have identified several limitations with BG's design. These shape our research efforts to extend BG to maintain it as a state of the art benchmark. Most are in the context of the novel features of BG that make it unique. Below, we describe these in turn, detailing BG's scalable request generation in Sect. 2, its closed emulation of socialites and an alternative open emulator in Sect. 3, its rating mechanism in Sect. 4, its validation phase in Sect. 5, and additional actions in Sect. 6. Section 7 provides our long term future research.

2   Scalability

BG employs a shared-nothing architecture and scales to a large number of nodes, preventing either the CPU, network, or memory resources of a single node from limiting its request generation rate. Its software architecture consists of one coordinator and N clients, termed BGCoord and BGClient, respectively. In our experiments with an 8 core CPU, a multi-threaded BGClient is able to utilize all cores fully as long as the client component of a data store does not suffer from the convoy phenomena [6] and the data store is able to process requests at the rate generated by BG. When the client component of a data store limits vertical scalability, as long as there is a sufficient amount of memory, one may execute multiple instances of BGClients on a single node to utilize all cores. BG scales horizontally by executing multiple BGClients across different nodes. BGCoord is responsible for initiating the BGClients, monitoring their progress, gathering their results at the end of an experiment, and aggregating the obtained results to compute the SoAR of a data store.

Once the BGClient instances are started, they generate requests independently with no synchronization. This is made possible using the following two concepts.

First, a BGClient implements a decentralized partitioning strategy that declusters a benchmark social graph into N disjoint sub-graphs where N is the number of BGClients. A BGClient is assigned a sub-graph to generate requests referencing members of its assigned sub-graph only. While the data store is not aware of this partitioning, the data generated and stored in the data store does correspond to the N disjoint graphs. One may conceptualize each sub-graph as a province whose citizens may perform BG's actions with one another only. This means citizens of different provinces may not view one another's profile or become friends with one another.

Second, BG employs a novel decentralized implementation of the Zipfian distribution, named D-Zipfian [3,24], that ensures the distribution of requests to the different members is independent of N. Thus, the distribution of access with one node is the same as that with several nodes. D-Zipfian in combination with partitioning of the social graph enables BG to utilize N nodes to generate requests without requiring coordination until the end of the experiment, see [2–4] for details.

While BG scales to a large numbers of nodes, its two concepts may fail to evaluate some data stores objectively. As an example, consider the architecture of Fig. 2 where an application is extended with a cache such as KOSAR or EhCache [12]. This caching framework consists of a KOSAR coordinator that maintains which application server has cached a copy of a data item in its KOSAR JDBC wrapper, KOSAR-Client for short. When one application server updates a copy of the data item, its KOSAR-Client informs the KOSAR coordinator of the impacted data item. In turn, the KOSAR coordinator invalidates a copy of this data item that resides in the KOSAR-Client of other application servers. With a skewed pattern of access to members and a workload that exhibits a low read to write ratio, a centralized KOSAR coordinator may become the bottleneck and dictate the overall system performance. The aforementioned

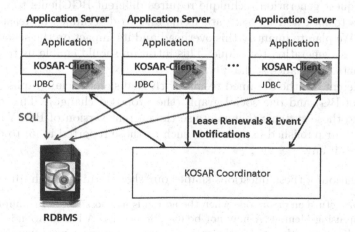

Fig. 2. A data intensive architecture using KOSAR.

two concepts employed by BG fail to cause the formation of such a bottleneck. To elaborate, each application server references data items that are unique to itself since its assigned sub-graph is unique and independent of the other sub-graphs. Hence, once an application server updates a cached data item, BG does not exercise the KOSAR coordinator informing KOSAR-Client of another application server.

To address the above limitation, we are extending BG to employ N BGClients with one social graph. The key concept is to hash partition members and resources across the N BGClients. Each BGClient is aware of the hash function and employs the original Zipfian distribution (instead of D-Zipfian) to generate member ids. When a BGClient BGC_i references a data item that does not belong to its assigned partition, it contacts the BGClient that owns the referenced data (say BGC_j) to lock that data item for exclusive use by BGC_i and to determine if its intended action is possible. BGC_j grants the lock request if there is no existing lock on the referenced data item and the action is possible, enabling BGC_i to proceed to generate a request with the identified data item to the data store. Once the request is serviced, BGC_i contacts BGC_j to release the exclusive lock on the referenced data item to make it available for use by other BGClients. This design raises the following interesting questions:

- When BGC_j fails to grant an exclusive lock[2] to the referenced data item due to an existing lock, how should the framework handle the conflict? Three possibilities are as follows. First, it may block BGC_i until the referenced data item becomes available. Second, it may return an error to BGC_i to generate a different member/resource id and try again. Third, it may simply abort this action and generate a new action all together. We intend to quantify the tradeoff associated with these three possibilities and their impact on both the distribution of requests and the benchmarking framework.
- What is the scalability characteristic of the proposed technique? The proposed request generation technique requires different BGClients to exchange messages to lock and unlock data items and to determine the feasibility of actions. We plan to quantify this overhead and its impact on the scalability of this request generation technique. This intuition should enable us to propose refinements to enhance scalability.
- How different are the obtained results with N disjoint social graphs (current version of BG) and one social graph (the proposed change)? This question applies to those systems that may use the current version of BG. We intend to repeat our published experiments such as those reported in [5] to quantify differences if any.

An investigation of these questions shapes our short term research direction.

---

[2] BGC_j may return an error code when the action is not possible. For example, Thaw Friendship using Member A may not be feasible because A has no friends. In these cases, BGC_i may either abort the action or may reference a new member for the same action.

3   Closed Versus Open

BGClients generate requests using a fixed number of threads T. Each thread emulates a random member of a social networking site performing one of the eleven actions. The randomly selected member is conditioned using the D-Zipfian distribution. This is termed a closed emulation model because a thread does not emulate a new member generating a new action until its emulation of a current member completes. This model may include a think time between emulation of different members issuing actions. Historically, this is a model[3] of a financial institution with a fixed number of tellers (ATM machines) with T concurrent customers (threads) performing financial transactions simultaneously [13].

An open emulator is a more realistic model of a social networking site [17] (and web sites in general). With this model, the emulator generates requests based on a pre-specified arrival rate, λ. This model is depicted in Fig. 3 where a factory generates members who issue a social networking action independently. (A member who is performing an action is termed a *socialite*.) The factory does not wait for the data store to service a request issued by a socialite. Instead, it generates λ socialites issuing requests per unit of time using a distribution such as random, uniform, or Poisson. A Poisson distribution results in a pattern of requests that is bursty. This means λ is an average and the number of simultaneous socialites at an instance in time might be higher than λ.

While the open emulator is more realistic, its design and implementation requires a careful study. This is because today's data stores service requests at such a high rate that the emulator must support λ values in the order of a million without exhausting its CPU resources. In addition to the scalability discussions of Sect. 2, the emulator must generate requests in a burst consistent with the Poisson distribution. At the time of this writing, we are evaluating the feasibility of such an open emulator and its implementation in BG.

4   Rating Mechanism

BG *rates* a data store to compute its Social Action Rating (SoAR) and Socialites rating for processing a workload. A workload consists of a mix of the eleven actions (see Table 1 and its discussion in Sect. 1), an exponent for the D-Zipfian distribution to control its degree of skew when referencing members, and a pre-specified SLA, see Sect. 1 for an example SLA. The SoAR of a data store is the highest throughput provided by that data store for the specified workload. The Socialite rating of a data store is the maximum number of simultaneous socialites (threads) that may generate requests corresponding to the specified workload. The pre-specified SLA imposes constraints on the acceptable response times and the amount of unpredictable data to constrain the SoAR and Socialite ratings of a data store. Given several data stores, the data store with the highest SoAR and Socialite rating is the superior one.

---

[3] This model is a representative of a web server configured with a maximum number of threads.

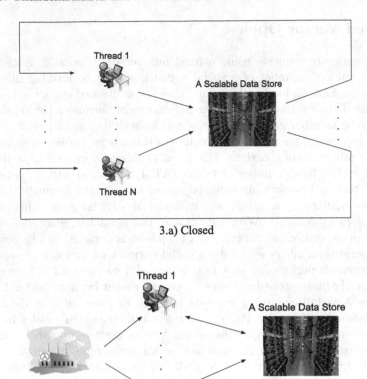

Fig. 3. Closed and open emulation of socialites issuing actions to a data store.

BG's rating process and its assumptions are detailed in [2]. Briefly, the rating process consists of a heuristic search that conducts multiple experiments. Each experiment uses the same workload specified by an experimentalist. With those workloads that impact the state of the database, the rating process might be required to re-load the database prior to each experiment. The repeat loading of a benchmark database may constitute a significant portion of the rating process. For example, the time to generate a one million member database with a data store requires 11 days [4]. If the rating process conducts ten experiments each 30 min in duration, the time to generate the database each time would require almost 4 months.

Three alternatives that re-generate the database expeditiously are detailed in [4]. One is the Database Image Loading (*DBIL*) technique that maintains the original image of a database and copies it as the current database in advance of each experiment to reduce the load time. The time to copy the one million member database is 30 min [4], reducing the time to conduct ten experiments to

11 days and 10 h. The 11 days incurred to create the database for the first time is a one time overhead. By maintaining the image and re-using it, the duration of rating of the data store with different workloads is reduced dramatically. For example, the time to conduct ten experiments is now 10 h. We refer the interested reader to [4] for a detailed description of DBIL and two other techniques.

A key research question is the duration of each conducted experiment. A possible answer is to employ the duration specified by the SLA, Δ. However, if an experimentalist select a high value for this input parameter, then the rating process may consume more time than necessary. It is desirable to run experiments for shorter durations than Δ to identify a region in the search space that establishes the true SoAR of a data store. Ideally, the duration of an experiment should be the smallest possible value, δ, that reflects the behavior of a data store as if the experiment was running for Δ time units. The ideal δ is both data store and workload dependent and can be analyzed in a pre-processing step, prior to the rating process. This step involves multiple experiments issuing the given workload against the data store to select the smallest duration that results in a *steady* system behavior defined as one whose resource utilization and observed throughput do not change in time. For such a system the recently observed behavior will continue to hold into the future.

5   Validation

A novel feature of BG is its ability to quantify the amount of unpredictable data (stale, inconsistent, erroneous) produced by a data store. A data store may produce unpredictable data for a variety of reasons. Examples include use of a weak consistency technique such as eventual [20,23] and use of a cache [12] in a manner that results in dirty reads [15] and inconsistent cache states [11].

BG is able to measure the amount of unpredictable data because it is a stateful benchmark that is aware of the initial state of a data item in the database and its updates to the database. It maintains the start and end of each action to enumerate the finite number of ways a read may overlap multiple concurrent write actions that reference the same data item. BG enumerates these to compute a range of possible values that should be observed by the read operation. If a data store produces a different value then it has produced unpredictable data. This process is named *validation*.

BG decouples generation of requests to quantify the performance of a data store from the validation phase, performing validation in an off-line manner. When generating requests, each BG thread generates a log record for each of its actions: read log records for read actions and write log records for write actions. These log records are written to separate files. One file for the read log records and a second file for the write log records.

The validation phase assumes in-memory data structures to compute the amount of unpredictable data as follows. First, it maintains an interval tree for (1) each member and write actions that impact her friendships, (2) each member and write actions that impact her pending friend invitations, and (3) each resource that is annotated with a write action. It constructs interval trees for a member/resource on demand as it reads the write log records in memory. The start and end time stamp of the write log records are indexed by an interval tree. Once the write log records are staged in memory, the validation phase retrieves the read log records. It employs the member id (resource id) and the action to identify the interval tree with the relevant write log records. Next, it uses the start and end time stamp of each read log record to enumerate the number of ways it overlaps with the different write actions, computing a range of valid values that should be observed by the read action. If the value observed by the read action (recorded in the log record) does not match one of the valid values then this read action has observed an unpredictable value.

The current implementation of the validation phase is fast when there is a sufficient amount of memory to stage the write log records in interval trees. This enables the validation phase to read the log files once to process both read and write log records in one pass. This multi-threaded process may utilizes multiple cores fully because (1) write log records are inserted into different interval trees based on their referenced member/resource id, and (2) different read log records may share and read the same interval tree simultaneously.

By performing validation in an off-line manner, BG reduces the number of nodes required to generate request to evaluate a data store. A key limitation of the current validation technique is that it may exhaust the available physical memory, causing the operating system to exhibit a thrashing behavior that results in an unacceptably long validation process. This is specially true with high throughput multi-node data stores and cache augmented data stores such as KOSAR that process requests in the order of millions of actions per second. We intend to extend the validation phase to analyze the size of files produced during an experiment to estimate the amount of required memory to ensure it does not exhaust the physical memory. In passing, it is interesting to note that the log records pertaining to each unique member can be processed independently to identify unpredictable data. Hence, a MapReduce [8] framework such as Hadoop is effective in implementing the validation phase.

6   Additional Actions

The eleven actions of BG are a good start to evaluate a data store. However, there are other actions that are common to many social networking sites that we intend to abstract. An implementation of these would extend BG with new actions. Here, we focus on two new actions to be released soon, namely, Share Resource (SR) and Retrieve Feed (RF). Both are in support of feed following [1,18,19]. This action is supported by sites such as Google+, Twitter, Facebook, My Yahoo and others. It enables users to create personalized feed by selecting one or more event streams they wish to follow.

Figure 4 shows the high level (incomplete) ER diagram for feed following. While the conceptual model looks complex, it is based on the concept of aggregation that establishes the many-to-many relationship between producers and

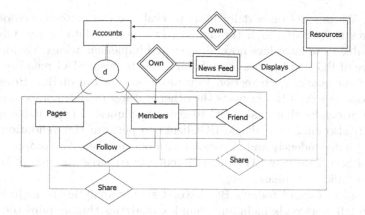

Fig. 4. Conceptual design of BG's feed following (Color figure online).

consumers of news feeds. The producers are a specialization of the Accounts entity set, identifying Pages and Members as two different categories because pages may have a significantly larger number of followers[4] and do not perform some of the actions performed by members. Examples of pages are celebrity fan pages and company and brand pages among the others. Resources such as images and tweets posted by a page are shared with their followers. This is represented by the red rectangle that aggregates the follow relationship between members and pages as an entity that participates in the "Share" relationship (red line) with the Resources entity set. Similarly, resources posted by the members are shared with their circle of friends which are also members. This is denoted by the green aggregation that participates in the "Share" relationship with the Resources entity set. Each member owns a "News feed" that contains resources shared by those whom they are following.

In the near future, we intend to investigate alternative techniques to implement feed following; see [19] for a pull, push, and a hybrid of these two techniques. We are also exploring techniques that compute an approximate feed. We intend to use BG to quantify both the scalability of these techniques and the performance gain of the approximate technique relative to its amount of unpredictable data.

7 Conclusions and Future Research

Social networks are emerging in diverse applications that strive to provide a sense of community for their users. These diverse applications range from financial web sites such as online trading system to academic institutions. BG is a benchmark to evaluate the performance of a data store for processing interactive social networking actions such as viewing a member's profile, extending a friendship request to a member, accepting a friendship request, and others as shown

---

[4] As of November 12, 2013, Katy Perry had more than forty million Twitter followers.

in Table 1. We use BG on a daily basis to evaluate the performance of novel architectures that enable high throughput, low latency data stores. Obtained insights enable us to introduce novel designs and implementations. We identified key features of BG and how we are refining these to ensure BG remains a state of the art benchmark. Each section focuses on one feature of BG. However, a change in one aspect of BG impacts the other features of BG. For example, the two design principles that enable BG to generate requests in a scalable manner (see Sect. 2) also enables multiple BGClients to perform the validation phase (see Sect. 5) independently and in parallel with one another. The proposed modifications of Sect. 2 on how BG generates requests requires a revisit of how BG performs its validation phase.

We intend to extend today's BG beyond simple analytics to include additional Web 2.0 workloads including complex analytics that require the use of machine learning algorithms. An example is a recommender system that analyzes the social graph and attribute values to suggest friends for a member. A challenge here is to generate the social graph in a manner where the output of the recommendation system is known, extending BG with metrics such as precision and recall. The graph database may include annotations on the edges of the graph with attribute values. For example, the friendship relationship between two members A and B might be tagged with a family relationships (such as daughter), number of likes given by Member A to postings by B's friends, number of comments posted by B's friends on A's resources, and others. We may change how the value of textual attributes for a comment are generated in a manner that is more realistic using techniques such as those utilized by the TREC benchmark. This may evaluate alternative (1) natural language processing and (2) information retrieval techniques in the context a recommender system for a social graph.

Finally, we are investigating the viability of a *Benchmark Generator*, BG+, that inputs an abstraction of an application, its actions and their dependencies, metrics of interest to be quantified, and a control parameter. Its output is a benchmark specific to that application. This is appropriate for those applications with diverse use cases such as data sciences [14,22]. In essence BG+ is an extensible toolkit that is configured with its input to enable a data scientist to develop a benchmark for their activities rapidly. Similar to BG and YCSB, it would expose the implementation of the database schema and the abstracted actions to be implemented by the data scientists. This enables BG+ to support diverse types of data models such as structured (relational), un-structured, JSON, extensible, images, audio and video as input. The control parameter may manipulate factors such as noise in the data (similar to today's system load controlled by parameter T) to analyze the behavior of an algorithm. This would enable BG+ to output a rating mechanism that computes a single value (a maxima such as SoAR) by manipulating the value of the control parameter.

References

1. Bai, X., Junqueira, F.P., Silberstein, A.: Cache refreshing for online social news feeds. In: CIKM, pp. 787–792 (2013)
2. Barahmand, S., Ghandeharizadeh, S.: BG: a benchmark to evaluate interactive social networking actions. In: CIDR, Jan 2013
3. Barahmand, S., Ghandeharizadeh, S.: D-Zipfian: a decentralized implementation of Zipfian. In: ACM SIGMOD DBTest Workshop (2013)
4. Barahmand, S., Ghandeharizadeh, S.: Expedited benchmarking of social networking actions with agile data load techniques. In: CIKM (2013)
5. Barahmand, S., Ghandeharizadeh, S., Yap, J.: A comparison of two physical data designs for interactive social networking actions. In: CIKM (2013)
6. Blasgen, M.W., Gray, J., Mitoma, M.F., Price, T.G.: The convoy phenomenon. Oper. Syst. Rev. **13**(2), 20–25 (1979)
7. Cattell, R.: Scalable SQL and NoSQL data stores. SIGMOD Rec. **39**, 12–27 (2011)
8. Dean, J., Ghemawat, S.: MapReduce: simplified data processing on large clusters. In: Symposium on Operating Systems Design and Implementation, vol. 6 (2004)
9. Nishtala, R., et al.: Scaling memcache at Facebook. In: NSDI (2013)
10. Floratou, A., Teletria, N., DeWitt, D.J., Patel, J.M., Zhang, D.: Can the elephants handle the NoSQL onslaught? In: VLDB (2012)
11. Ghandeharizadeh, S., Yap, J.: Gumball: a race condition prevention technique for cache augmented SQL database management systems. In: ACM SIGMOD DBSocial Workshop (2012)
12. Ghandeharizadeh, S., Yap, J.: Cache augmented database management systems. In: ACM SIGMOD DBSocial Workshop, June 2013
13. Gray, J., Reuter, A.: Transaction Processing: Concepts and Techniques, pp. 677–680. Morgan Kaufmann, San Francisco (1993)
14. Greenberg, C.: Overview of the NIST data science evaluation and metrology plans. In: Data Science Symposium, NIST, 4–5 Mar 2014
15. Gupta, P., Zeldovich, N., Madden, S.: A trigger-based middleware cache for ORMs. In: Middleware (2011)
16. Patterson, D.: For better or worse, benchmarks shape a field. Commun. ACM **55**, 104 (2012)
17. Schroeder, B., Wierman, A., Harchol-Balter, M.: Open versus closed: a cautionary tale. In: NSDI (2006)
18. Silberstein, A., Machanavajjhala, A., Ramakrishnan, R.: Feed following: the big data challenge in social applications. In: DBSocial, pp. 1–6 (2011)
19. Silberstein, A., Terrace, J., Cooper, B.F., Ramakrishnan, R.: Feeding frenzy: selectively materializing users event feeds. In: SIGMOD Conference, pp. 831–842 (2010)
20. Stonebraker, M.: Errors in database systems, eventual consistency, and the CAP theorem. Commun. ACM. BLOG@ACM, Apr 2010
21. Stonebraker, M., Cattell, R.: 10 rules for scalable performance in simple operation datastores. Commun. ACM **54**, 72–80 (2011)
22. Talukder, A.: Overview of the NIST data science program. In: Data Science Symposium, NIST, 4–5 Mar 2014
23. Vogels, W.: Eventually consistent. Commun. ACM **52**(1), 40–45 (2009)
24. Yap, J., Ghandeharizadeh, S., Barahmand, S.: An analysis of BGs implementation of the Zipfian distribution. USC DBLAB technical report 2013-02 (2013). http://dblab.usc.edu/Users/papers/zipf.pdf

A Micro-benchmark Suite for Evaluating Hadoop RPC on High-Performance Networks

Xiaoyi Lu$^{(\boxtimes)}$, Md. Wasi-ur-Rahman, Nusrat Sharmin Islam,
and Dhabaleswar K. (DK) Panda

Database Laboratory, Computer Science Department,
USC, Los Angeles, CA 90089-0781, USA
{luxi,rahmanmd,islamn,panda}@cse.ohio-state.edu

Abstract. Hadoop Remote Procedure Call (RPC) is increasingly being used with other data-center middlewares such as MapReduce, HDFS, and HBase in many data-centers (e.g. Facebook, Yahoo!) because of its simplicity, productivity, and high performance. For RPC systems, achieving low latency and high throughput is critical. However, a standardized benchmark suite that focuses on helping users evaluate the performance of standalone Hadoop RPC is lacking in current Apache Hadoop distribution. In this paper, we design and develop a micro-benchmark suite that can be used to evaluate the performance of Hadoop RPC in terms of latency and throughput with different data types. We show how this benchmark suite can be used to evaluate the performance of Hadoop RPC over different networks/protocols and parameter configurations on modern clusters.

Keywords: Big Data · Hadoop RPC · Micro-benchmarks · High-performance networks

1   Introduction

Hadoop [27] is one of the most popular open-source frameworks that can handle massive data analysis, storage, and query. It supports the MapReduce [11] programming model, Hadoop Distributed FileSystem (HDFS) [25], Hadoop database, HBase [6], and so on. Hadoop is being used in many data-center applications at companies such as Facebook and Yahoo! because of its scalability, fault-tolerance, and productivity.

Hadoop Remote Procedure Call (RPC) is the fundamental communication mechanism in Hadoop [29]. It is mainly used for metadata exchange; and, is used in all the Hadoop components such as MapReduce, HDFS, and HBase. In MapReduce, it is used for control signaling, which manages compute nodes and tracks system status. Similarly in HDFS, it is used for communication between data nodes and name node for efficient data management operations; such as,

---

This research is supported in part by National Science Foundation grants #OCI-0926691, #OCI-1148371 and #CCF-1213084.

© Springer International Publishing Switzerland 2014
T. Rabl et al. (Eds.): WBDB 2013, LNCS 8585, pp. 32–42, 2014.
DOI: 10.1007/978-3-319-10596-3_3

getting block info and creating data blocks. Further, it is used as the core communication scheme for HBase Put/Get database operations.

Building up Hadoop RPC protocol is very simple; users just need to define the interfaces and remote procedures. Hadoop RPC internally takes care of the serialization/deserialization and the proper invocation of remote procedures. Interface arguments and return types can be primitives such as `String`, `Double`, or `Array`, etc. They can also have a specialized serialization format in Hadoop based on *Writable* [29].

As the core component in the Hadoop software stack, the RPC performance, in terms of latency and throughput, dramatically influences the performance of middleware and components in the upper layer. The performance of Hadoop RPC is determined by many factors such as network configurations in cluster, controllable parameters in software (e.g. number of handler), data types, and so on. In order to get optimal performance, it is better to tune these factors based on cluster and workload characteristics. Adopting a standardized performance benchmark suite to evaluate the performance metrics in different kinds of configurations is a good choice for Hadoop users. For Hadoop developers, this kind of benchmark suite is also helpful to evaluate the performance of new designs. However, currently we lack a standardized benchmark suite that focuses on helping users evaluate the performance of standalone Hadoop RPC in the current Apache Hadoop distribution. The following is the basic motivation of this paper: *Can we design and implement a simple and standardized benchmark suite to let users and developers in the Big Data community evaluate, understand, and optimize the Hadoop RPC performance over a range of networks/protocols?*

In this paper, we design and implement a comprehensive micro-benchmark suite to evaluate the performance of standalone Hadoop RPC. We provide options for varying different benchmark-level parameters such as data type, minimal/maximum payload size, number of concurrent clients, round number of each test, etc. Our benchmark suite can also dynamically set the Hadoop RPC configuration parameters, like numbers of concurrent handlers in server side, etc. We display the configuration parameters for each test, as part of the benchmark output and present different statistics like min, max, and average, for the results.

This paper makes the following key contributions:

1. Design and implementation of a micro-benchmark suite to evaluate the performance of standalone Hadoop RPC.
2. A set of standard benchmarks to measure the latency and throughput of Hadoop RPC with different data types.
3. Illustration of the performance results of Hadoop RPC using these benchmarks over different networks/protocols (1 GigE/10 GigE/IPoIB).
4. A case study of enhancing Hadoop RPC design over native InfiniBand with the help of these benchmarks.

The rest of the paper is organized as follows. In Sect. 2, we discuss related work in the field. We present our design considerations for the benchmark suite in Sect. 3 and benchmarks in Sect. 4. In Sect. 5, we show our performance results. Section 6 shows a case study. Finally, we conclude the paper in Sect. 7.

2   Related Work

The concept of Remote Procedure Call (RPC) goes back to at least the 1980s. Birrell and Nelson [9] first presented their implementation of a high-performance RPC system. After that, many kinds of RPC implementations have been popularly used in the distributed computing area [4]. Recently, many open-source projects are proposed to provide high-performance RPC systems, such as Avro [26], Thrift [7], Protocol buffers [2], and so on. The basic motivation of these systems is to provide a more flexible data representation and a better data serialization and de-serialization mechanism. Because of simplicity, productivity, and high performance, Hadoop RPC has been already widely used with MapReduce, HDFS, and HBase in many data-centers. However, without the availability of a standardized suite of micro-benchmarks to evaluate the Hadoop RPC performance in current Hadoop, users and developers are not able to analyze performance characteristics of Hadoop RPC on different platforms.

Many other related benchmarks have been provided in Cloud Computing and Big Data areas. The study in [13] proposed a benchmark suite for comparing the performance of SOAP engine implementations under a wide range of representative use cases. The authors in [20,21] discussed performance characteristics and programming efforts in different languages for SOAP services by their designed benchmarks. MRBench [17] provides micro-benchmarks in the form of MapReduce jobs of TPC-H [5]. MRBS [24] is a benchmark suite that evaluates the dependability of MapReduce systems. It provides five benchmarks for several application domains and a wide range of execution scenarios. The authors of HiBench [14] have extended the DFSIO program to compute the aggregated throughput by disabling the speculative execution of the MapReduce framework. HiBench also evaluates Hadoop in terms of system resource utilization (e.g. CPU, memory). The micro-benchmark suite designed in [15] helps detailed profiling and performance characterization of various HDFS operations. MalStone [8] is a benchmark suite designed to measure the performance of cloud computing middleware when building data mining models. Yahoo! Cloud Serving Benchmark (YCSB) [10] is a set of benchmarks for performance evaluations of key/value-pair and cloud data serving systems. YCSB++ [22] further extends YCSB to improve performance understanding and debugging. BigData-Bench [28], a benchmark suite for Big Data Computing, covers typical Internet service workloads and provides representative data sets and data generation tools. It also provides different implementations for various Big Data processing systems [1,18].

Our proposed benchmark suite addresses the shortcoming in the Big Data community for evaluating Hadoop RPC performance. Our benchmark suite does not need to launch the whole Hadoop and can be used in a simple manner to evaluate the performance of the standalone Hadoop RPC engine. As a result, our benchmark suite can be used to carry out performance comparison of Hadoop RPC for different network, protocol, and parameter configurations on modern clusters.

3 Design Considerations

The performance of RPC systems is usually measured by the metrics of latency and throughput. The performance of Hadoop RPC is influenced by the factors such as underlying network (or communication protocol), configuration parameters (e.g. handler threads) of RPC, and CPU utilization. We consider these aspects described below when we design the benchmark suite.

Network: The performance of Hadoop RPC operations is significantly influenced by the underlying interconnect or protocol. During a RPC call, the serialized data is transferred from one node to another. Therefore, high performance networks can speed up the transmission of messages. Our benchmark suite can be used to evaluate the Hadoop RPC performance over different network types, and it can help users to understand the interaction between network characteristics and Hadoop RPC performance.

Configuration Parameters: The performance of Hadoop RPC largely depends on various configuration parameters. From the Hadoop RPC usage perspective, current Hadoop RPC supports users to tune different number of processing handlers in the RPC engine. In our benchmarks, we provide options to set this kind of configuration parameter at runtime by users. If no values are provided from the user level, the benchmark should run with the parameters by default. Besides, as a benchmark suite, we should support several benchmark-level parameters to help users do experiments as efficiently as possible.

Data Types: Many researches (e.g. [12,13]) have discussions with serialization and de-serialization issues of different data types in RPC systems. A comprehensive RPC benchmark suite should have this type of support for users to test with different kinds of data types with very simple configurations. Our proposed benchmark suite supports different data types in Hadoop that are based on the `Writable` interface.

CPU Utilization: In fact, CPU utilization is a trade-off of the design in modern RPC systems. On the one hand, RPC system will gain better performance if it can take up as much as possible CPU cycles. However, on the other hand, the RPC system is often used as an internal component in a distributed computing system. If the RPC engine takes up lots of CPU cycles, then other components in the same system would be influenced. Therefore, the benchmark suite for RPC systems should also support monitoring CPU utilization.

4 Benchmarks for Hadoop RPC

In this study, we design and implement two major benchmarks. Both of them can be used to test with different networks.

Latency: This client-server benchmark measures the ping-pong latency of Hadoop RPC communication for different message sizes and different data types in Hadoop. The user can vary the data type, minimum and maximum message sizes, number of iterations for the RPC client, and the number of handler threads for the RPC server. We can also support users to understand the detail of our benchmark by enabling the option of `verbose`. The detailed parameter list is shown in Table 1.

Table 1. Parameter list of Hadoop RPC latency benchmark

Component	Network address	Port	Data type	Min Msg Size	Max Msg Size	No. of iterations	Handlers	Verbose
lat_client	✓	✓	✓	✓	✓	✓		✓
lat_server	✓	✓					✓	✓

Table 2. Parameter list of Hadoop RPC throughput benchmark

Component	Network address	Port	Data type	Min Msg Size	Max Msg Size	No. of iterations	No. of clients	Handlers	Verbose
thr_client	✓	✓	✓	✓	✓	✓			✓
thr_server	✓	✓			✓		✓	✓	✓

Throughput: This benchmark measures the throughput for RPC communication for different message sizes and different data types in Hadoop. This is a single-server, multi-client benchmark. The user can vary the data type, minimum and maximum message sizes, number of iterations for the RPC client, and the number of handler threads for the RPC server. For each particular message size, every client calculates the throughput and sends it to the server. The server gets the throughput values from all the clients and thus calculates the minimum, average, and maximum throughput of them. This kind of detailed description and performance values of the benchmark will be printed by enabling the `verbose` option. By default, the server will report the aggregated throughput of all the clients for different message sizes. The detailed parameter configuration is shown in Table 2. Our server component utilizes "No. of Clients" to implement a simple barrier mechanism by an atomic counter that makes all the clients start their invocations almost at the same time. Besides, the `thr_server` program uses the "Max Msg Size" parameter to automatically finalize itself after all configured tests are finished. These strategies help users in conducting experiments. Furthermore, we supply a configuration option in our scripts to enable the resource utilization monitoring (e.g. CPU Utilization), which can be used to analyze the performance characteristics.

5   Performance Evaluation

5.1   Experimental Setup

Our testbed consists of nine nodes with Intel Westmere series of dual quad-core processors operating at 2.67 GHz with a 160 GB hard disk. Each node is equipped with a 1 GigE interface adapter and MT26428 QDR ConnectX HCAs (32 Gbps data rate) with PCI-Ex Gen2 interfaces. The nodes are interconnected by using a 171-port Mellanox QDR switch. Each node runs Enterprise Linux Server release 6.1 (Santiago) at kernel version 2.6.32-131 with OpenFabrics version 1.5.3. Each of these nodes has 24 GB of RAM. These nodes are also equipped with a Net-Effect NE020 10 Gb Accelerated Ethernet Adapter (iWARP RNIC) each. The 10 GigE cards are connected using a Fulcrum Focalpoint 10 GigE switch. This cluster is less than four years old and represents the leading-edge technologies used in commodity clusters for Big Data applications. In all our experiments, the Hadoop RPC server and clients run exclusively on different nodes. The version of Hadoop RPC is 1.2.1 and the JDK version is 1.7.

5.2   Micro-benchmark Results

In this section, we show some performance results using our micro-benchmarks presented in Sect. 4. We evaluate these benchmarks for three different networks/ protocols: 1 GigE, 10 GigE, and InfiniBand (32 Gbps)/IPoIB. All experiments with small message sizes are averaged over 20,000 iterations by considering the first 50 % as warm-up. For large message sizes, the number of iterations is reduced to 5,000.

In the first set of experiments, we evaluate the latency benchmark for BytesWritable and compare the performance in different interconnects. We vary message sizes from 1 Byte to 4 KB and these results are presented in Fig. 1(a). For larger message sizes, we conduct similar experiments and these results are represented in Fig. 1(b).

Both for small and large payload sizes, we can see from Fig. 1 that, the RPC latency decreases if the underlying interconnect is changed to IPoIB or 10 GigE from 1 GigE. With 10 GigE interconnect, we observe better latency than IPoIB for small message sizes. For large message sizes, IPoIB performs better than 10 GigE. As an example, IPoIB achieves 27 % gain over 10 GigE for the 64 MB message size, whereas it performs worse by 0.66 % over 10 GigE for the 4 KB message size.

Figure 2(a) and (b) show similar performance comparisons for RPC latency with the data type of Text.

For the second set of experiments, we use our benchmarks to compare the throughput in different network interconnects. For these experiments, we use BytesWritable for RPC communication. In Fig. 3(a), we show throughput comparison with seven handlers in the RPC server. We run seven clients for this experiment and each client runs on a separate node. For Fig. 3(b), we increase the handler count in server to 16 for the same number of clients. Here also

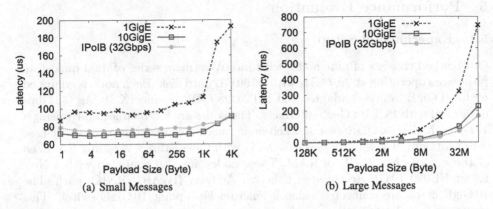

Fig. 1. RPC latency for `BytesWritable`

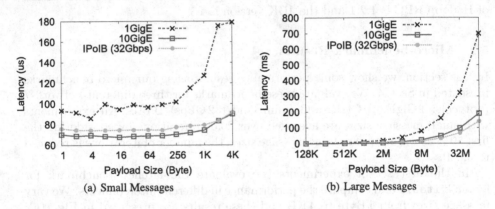

Fig. 2. RPC latency for `Text`

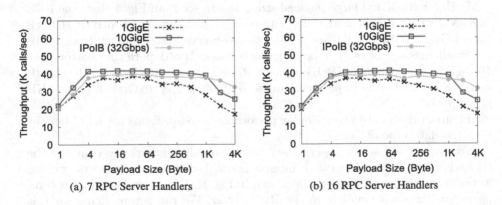

Fig. 3. RPC throughput for `BytesWritable`

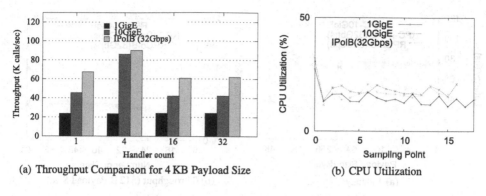

(a) Throughput Comparison for 4 KB Payload Size   (b) CPU Utilization

Fig. 4. Throughput and CPU utilization

we observe similar trends. IPoIB performs better than 10 GigE as message size is increased and at 4 KB, the improvement goes up to 26 % for seven handler threads. For small message sizes, 10 GigE still performs better than IPoIB by an average margin of 5–6 %.

We conduct similar experiments for more number of clients. Figure 4(a) illustrates this scenario. In this experiment, we vary the number of handlers and keep the payload size fixed at 4 KB. We run 42 clients for this experiment and observe the performance comparison among different interconnects. Figure 4(b) shows CPU utilization for the experiment with 16 handlers in server and seven clients. This figure shows that our benchmark suite can be easily used to monitor resource utilization.

6   A Case Study: Enhance Hadoop RPC Design Over Native InfiniBand

In our previous work [19], we have designed and implemented a high-performance Hadoop RPC system with RDMA over native InfiniBand, which is called RPCoIB and is public available through the RDMA for Apache Hadoop project [3,16,19,23]. During the design and optimization process, we find our proposed standalone Hadoop RPC benchmark suite in this paper is really helpful to evaluate the performance of alternative designs. Based on tested numbers, we can easily decide which design or optimization technique should be used. We can also easily choose different internal parameters based on the performance and resource utilization numbers outputted by the benchmark suite.

As illustrations, this section shows the performance comparisons of RPCoIB (available with RDMA for Apache Hadoop v0.9.9) with default Hadoop RPC over 10 GigE and IPoIB by using our proposed benchmarks. Figure 5 shows the results of latency and throughput. From these figures, we can see that compared with the RPCoIB design, the default Hadoop RPC over high-performance networks (10 GigE and IPoIB) can not achieve optimal performance. As shown in Fig. 5(a), when the payload sizes vary from 1 byte to 4 KB, RPCoIB can show

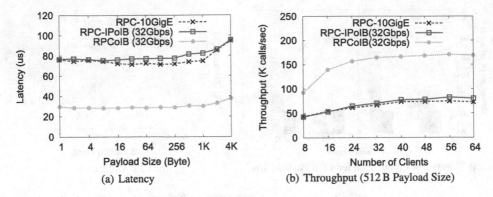

Fig. 5. Performance evaluation of different Hadoop RPC designs on 10 GigE and Infini-Band networks (Data type: `BytesWritable`)

59 %–62 % and 60 %–63 % improvements as compared with the performance of default Hadoop RPC on 10 GigE and IPoIB (32 Gbps), respectively. If we compare the RPCoIB performance with default Hadoop RPC running on low-speed network (e.g. 1 GigE), we can see that the RPCoIB design has about 1.92-4.29x performance speedup. In order to make the figures clearer, we do not show the numbers of low-speed network cases. From Fig. 5(b), we can see that the peak throughput of RPCoIB outperforms the peak performance of default Hadoop RPC on 10 GigE by 129 % and IPoIB by 107 %, respectively.

7   Conclusion and Future Work

This paper proposed a comprehensive and easy-to-use micro-benchmark suite for evaluating performance of standalone Hadoop RPC. It can support users to evaluate Hadoop RPC performance in terms of latency and throughput with different data types over different networks. It also provides flexibility to users for varying different benchmark configurations. Our micro-benchmark suite can help users understand the performance characteristics of Hadoop RPC by printing detailed information and monitoring resource utilization in a controllable manner. As an illustration, we have presented performance results of our benchmarks over different interconnects on a modern cluster. Our proposed micro-benchmark suite can also help developers to enhance the RPC engine design. Through our case study, we have shown that with the help of our micro-benchmark suite, a new RPC system design over native InfiniBand is proposed, which can achieve much better performance than that of default Hadoop RPC over high-performance networks.

We will continuously update our benchmark suite to help users to make the performance comparison among Hadoop Writable RPC, Avro, Thrift, and Protocol buffers. We plan to make this micro-benchmark suite available to the Big Data community via an open-source release in future. We hope our proposed benchmarks can be used as standardized benchmarks to beat the Hadoop RPC performance record in the Big Data community.

References

1. BigDataBench : A Big Data Benchmark Suite. http://prof.ict.ac.cn/BigDataBench
2. Protocol buffers. http://code.google.com/p/protobuf/
3. RDMA for Apache Hadoop. http://hadoop-rdma.cse.ohio-state.edu
4. Remote Rrocedure Call. http://en.wikipedia.org/wiki/Remote_procedure_call
5. TPC Benchmark H - Standard Speci cation. http://www.tpc.org/tpch
6. Apache HBase: http://hbase.apache.org/
7. Apache Thrift: http://thrift.apache.org/
8. Bennett, C., Grossman, R.L., Locke, D., Seidman, J., Vejcik, S.: Malstone: towards a benchmark for analytics on large data clouds. In: Proceedings of the 16th ACM SIGKDD International Conference on Knowledge Discovery and Data Mining, KDD, New York, NY, USA (2010)
9. Birrel, A.D., Nelson, B.J.: Implementing remote procedure calls. ACM Trans. Comput, Syst. **2**(1), 39–59 (1984)
10. Cooper, B.F., Silberstein, A., Tam, E., Ramakrishnan, R., Sears, R.: Benchmarking cloud serving systems with YCSB. In: Proceedings of the 1st ACM Symposium on Cloud Computing, SoCC, New York, NY, USA (2010)
11. Dean, J., Ghemawat, S.: MapReduce: simplified data processing on large clusters. In: Proceedings of the 6th Conference on Symposium on Opearting Systems Design and Implementation, OSDI, Berkeley, CA, USA (2004)
12. Ghazaleh, N.A., Lewis, M.J.: Differential deserialization for optimized SOAP performance. In: Proceedings of the 2005 ACM/IEEE Conference on Supercomputing, SC, Washington, DC, USA (2005)
13. Head, M.R., Govindaraju, M., Slominski, A., Liu, P., Abu-Ghazaleh, N., van Engelen, R., Chiu, K., Lewis, M.J.: A benchmark suite for SOAP-based communication in grid Web services. In: Proceedings of the 2005 ACM/IEEE Conference on Supercomputing, SC, Washington, DC, USA (2005)
14. Huang, S., Huang, J., Dai, J., Xie, T., Huang, B.: The HiBench benchmark suite: characterization of the MapReduce-based data analysis. In: Proceedings of the 26th International Conference on Data Engineering Workshops, ICDEW, Long Beach, CA, USA (2010)
15. Islam, N.S., Lu, X., Wasi-ur-Rahman, M., Jose, J., Panda, D.K.D.K.: A microbenchmark suite for evaluating HDFS operations on modern clusters. In: Rabl, T., Poess, M., Baru, C., Jacobsen, H.-A. (eds.) WBDB 2012. LNCS, vol. 8163, pp. 129–147. Springer, Heidelberg (2014)
16. Islam, N.S., Rahman, M.W., Jose, J., Rajachandrasekar, R., Wang, H., Subramoni, H., Murthy, C., Panda, D.K.: High performance RDMA-based design of HDFS over InfiniBand. In: Proceedings of the International Conference for High Performance Computing, Networking, Storage and Analysis, SC (2012)
17. Kim, K., Jeon, K., Han, H., gyu Kim, S., Jung, H., Yeom, H.: MRBench: a benchmark for MapReduce framework. In: Proceedings of the IEEE 14th International Conference on Parallel and Distributed Systems, ICPADS, Melbourne, Victoria, Australia (2008)
18. Liang, F., Feng, C., Lu, X., Xu, Z.: Performance benefits of DataMPI: a case study with BigDataBench. In: The 4th Workshop on Big Data Benchmarks, Performance Optimization, and Emerging Hardware, BPOE-4, Salt lake city, Utah (2014)
19. Lu, X., Islam, N.S., Rahman, M.W., Jose, J., Subramoni, H., Wang, H., Panda, D.K.: High-performance design of Hadoop RPC with RDMA over InfiniBand. In: Proceedings of the IEEE 42th International Conference on Parallel Processing, ICPP, Lyon, France (2013)

20. Lu, X., Lin, J., Zou, Y., Peng, J., Liu, X., Zha, L.: Investigating, modeling, and ranking interface complexity of Web services on the World Wide Web. In: Proceedings of the 6th World Congress on Services, SERVICES-1, Miami, Florida (2010)
21. Lu, X., Zou, Y., Xiong, F., Lin, J., Zha, L.: ICOMC: invocation complexity of multi-language clients for classified Web services and its impact on large scale SOA applications. In: Proceedings of the International Conference on Parallel and Distributed Computing, Applications and Technologies, PDCAT, Hiroshima, Japan (2009)
22. Patil, S., Polte, M., Ren, K., Tantisiriroj, W., Xiao, L., López, J., Gibson, G., Fuchs, A., Rinaldi, B.: YCSB++: benchmarking and performance debugging advanced features in scalable table stores. In: Proceedings of the 2nd ACM Symposium on Cloud Computing, SoCC, New York, NY, USA(2011)
23. Rahman, M.W., Islam, N.S., Lu, X., Jose, J., Subramoni, H., Wang, H., Panda, D.K.: High-Performance RDMA-based design of Hadoop MapReduce over InfiniBand. In: Proceedings of the IEEE 27th International Symposium on Parallel and Distributed Processing Workshops and PhD Forum, IPDPSW, Washington, DC, USA (2013)
24. Sangroya, A., Serrano, D., Bouchenak, S.: MRBS: towards dependability benchmarking for Hadoop MapReduce. In: Proceedings of the 18th International Conference on Parallel Processing Workshops, Euro-Par, Aachen, Germany (2013)
25. Shvachko, K., Kuang, H., Radia, S., Chansler, R.: The Hadoop distributed file system. In: Proceedings of the IEEE 26th Symposium on Mass Storage Systems and Technologies, MSST, Incline Village, Nevada (2010)
26. The Apache Software Foundation: Apache Avro. http://avro.apache.org/
27. The Apache Software Foundation: Apache Hadoop. http://hadoop.apache.org
28. Wang, L., Zhan, J., Luo, C., Zhu, Y., Yang, Q., He, Y., Gao, W., Jia, Z., Shi, Y., Zhang, S., Zheng, C., Lu, G., Zhan, K., Li, X., Qiu, B.: BigDataBench: a big data benchmark suite from Internet services. In: Proceedings of the 20th IEEE International Symposium on High Performance Computer Architecture, HPCA, Orlando, Florida (2014)
29. White, T.: Hadoop: The Definitive Guide. O'Reilly Media Inc., Sebastopol (2010)

Experience from Hadoop Benchmarking with HiBench: From Micro-Benchmarks Toward End-to-End Pipelines

Lan Yi[✉] and Jinquan Dai

Intel Asia-Pacific Research and Development Ltd.,
Shanghai 200241, People's Republic of China
lan.yi@intel.com, jasdai@microsoft.com

Abstract. As Hadoop-based big data framework grows in pervasiveness and scale, realistically benchmarking Hadoop systems becomes critically important to the Hadoop community and industry. In this paper, we present our experience of Hadoop benchmarking with HiBench (an open source Hadoop benchmark suite widely used by Hadoop users), and introduce our recent work on advanced end-to-end ETL-recommendation pipelines based on our experience.

Keywords: Big data · Hadoop · Benchmark · ETL · Recommendation

1 Introduction

As any good engineer knows, "if you cannot measure it, you cannot improve it"; benchmarking is the quantitative foundation of any computer system design and research. Among all existing big data frameworks (e.g., traditional data warehouse framework, MapReduce framework, in-memory and realtime processing framework, streams/graph processing framework, etc.), Hadoop-based framework is now the de facto standard. As Hadoop-based big data framework grows in pervasiveness and scale, realistically benchmarking Hadoop systems becomes critically important to the Hadoop community and industry.

However, the lack of industry standard performance benchmark that is representative of typical Hadoop workloads has inhibited the enterprise adoption of Hadoop. We started our work on Hadoop benchmarking with many customers with HiBench [1] (a comprehensive and representative Hadoop benchmark suite consisting of both synthetic micro-benchmarks and real-world workloads) from 2008, and released it to open source [2] in 2012. In this paper, we present our experience of Hadoop benchmarking with HiBench, and introduce our recent work on advanced end-to-end ETL-recommendation pipelines based on our experience.

Jinquan Dai: This work was done when the author was working in Intel Asia-Pacific Research and Development Ltd.

© Springer International Publishing Switzerland 2014
T. Rabl et al. (Eds.): WBDB 2013, LNCS 8585, pp. 43–48, 2014.
DOI: 10.1007/978-3-319-10596-3_4

2   Experience with HiBench

2.1   Overview of HiBench

Currently HiBench contains ten workloads, classified into four categories, as shown in Table 1.

Table 1. HiBench Workloads

Category	Workload
Micro benchmarks	Sort
	WordCount
	TeraSort
	EnhancedDFSIO
Web search	Nutch Indexing
	PageRank
Machine learning	Bayesian Classification
	K-means Clustering
Analytical query	Hive Join
	Hive Aggregation

Micro Benchmarks. The Sort, WordCount and TeraSort programs contained in the Hadoop distribution are three popular micro-benchmarks widely used in the community, and therefore are directly included in HiBench without any modification. Both the Sort and WordCount programs are representative of a large subset of real-world MapReduce jobs – one transforming data from one representation to another, and another extracting a small amount of interesting data from a large data set.

We have also extended the DFSIO program contained in the Hadoop distribution to evaluate the aggregated bandwidth delivered by HDFS. The original DFSIO program only computes the average I/O rate and throughput of each map task, and it is not straightforward how to properly sum up the I/O rate or throughout if some map tasks are delayed, re-tried or speculatively executed. The Enhanced DFSIO workload included in HiBench computes the aggregated bandwidth by sampling the number of bytes read/written at fixed time intervals in each map task; then it computes the aggregated read/write throughput by all the map tasks during the reduce and post-processing stage using linear interpolations.

Web Search. The Nutch Indexing and Page Rank workloads are included in HiBench, because they are representative of one of the most significant uses of MapReduce - large-scale search indexing systems.

The Nutch Indexing benchmark tests the indexing sub-system in Nutch [3], a popular open source (Apache project) search engine. It uses the automatically generated Web data whose hyperlinks and words both follow the Zipfian distribution with corresponding parameters. The Web data are organized in the internal defined format of Nutch so that they can be directly consumed by Nutch Indexing. In addition, the PageRank workload adopts an open source Pegasus [4] implementation of the widely used PageRank algorithm in Web search engines. The code of inputFormat and outputFormat of Pegasus are modified to support sequenceFile format and compression options. The workload uses the automatically generated Web data whose hyperlinks follow the Zipfian distribution.

Machine Learning. The Bayesian Classification and K-means Clustering implementations contained in Mahout are included in HiBench, because they are representative of one of another important uses of MapReduce (i.e., large-scale machine learning).

The Bayesian Classification workload implements the trainer part of Naive Bayesian (a popular classification algorithm for knowledge discovery and data mining) based on Mahout 0.7, which is an open source (Apache project) machine learning library. The K-means Clustering workload implements K-means (a well-known clustering algorithm for knowledge discovery and data mining) based on Mahout 0.7. We have developed a random data generator using statistic distributions to generate the input for the K-means Clustering workload.

Analytic Query. The Join and Aggregation queries in the Hive performance benchmarks [5] are included in HiBench, because they are representative of another one of the most significant uses of MapReduce (i.e., OLAP-style analytical queries).

Both Hive Join and Aggregation queries are adapted from the query examples in Pavlo et al. [6]. They are intended to model complex analytic queries over structured (relational) tables – Hive Aggregation computes the sum of each group over a single read-only table, while Hive Join computes both average and sum for each group by joining two different tables. Their inputs are automatically generated Web data with hyperlinks following the Zipfian distribution and the randomly distributed Web logs.

2.2 Tradeoffs Between HiBench and Synthetic Trace-Based Approaches

An important class of Hadoop benchmarks available today (e.g., Gridmix3 [7] and SWIM [8]) is based on synthetic trace-based approaches, which (1) collect traces of the jobs in the Hadoop cluster over an extended period of time, (2) synthesize the workload based on the traces, and (3) execute the workloads via replaying the synthesized traces. The major issue here is that this approach is only as good as the model it uses to synthesize Hadoop jobs into traces, and it is almost impossible to design an appropriate model that precisely captures all the intrinsic properties of a Hadoop job.

For instance, in GridMix3, the trace is generated by mining the job execution history logs collected by the JobTracker. For each job, it records a lot of information such as job submission time, numbers of map/reduce tasks, numbers of bytes and records read/written by each task, etc. Then the framework replays the trace by running synthetic Hadoop jobs simulating the recorded information. However, with the above information, the synthetic jobs can only simulate the I/O load. To emulating other resource usage (e.g., CPU and memory), it also collects CPU and memory usages for each task; when replaying the traces, each task in the synthetic jobs launches a separate thread to emulate these resource usages. Unfortunately, this approach does not precisely capture all the intrinsic properties of a Hadoop job; for instance:

- *The numbers of tasks is not purely an intrinsic property of the Hadoop job* – it depends on both the job itself (e.g., how much data to process) and the underlying Hadoop framework (e.g., how to partition the data & computation). For instance, one common Hadoop performance issue is when the Hadoop job contains a lot of small map tasks, and one way to address this issue is to enhance the Hadoop framework to automatically samples the input data and generate reasonably large input data splits; however, this improvement cannot be evaluated by simply replaying the traces of the original job.
- *It cannot capture the actual computation performed by the tasks* – an important property of the Hadoop job. Without this information, Gridmix3 just conducts some random math operations (e.g., multiply, square root, etc.), which does not represent real world processing in a Hadoop job. For instance, Hive queries are often CPU bottlenecked (most likely due to serialization, compression, sorting, etc.), and a processer can significant speeds up these queries if it provide better support for Java serialization (e.g., new instructions or accelerators); however, this improvement cannot be evaluated by performing random math operations.

HiBench has taken a different approach – it provides a collection of real-world Hadoop applications that are both representative and diverse, so that users can understand their characteristics by running the actual application codes on the underlying Hadoop cluster. On the other hand, today it only contains workloads/kernels which are isolated components of end-to-end Hadoop application pipelines. To address this issue, we are currently working on an end-to-end ETL-Recommendation pipeline.

2.3 ETL-Recommendation Pipeline

The ETL-Recommendation pipeline is a step of HiBench towards end-to-end benchmark. It simulates the daily offline workloads of a recommendation engine based on a retailer's Hive data warehouse. The pipeline updates the structured Web sales data and the unstructured Web logs data, and then recalculates the up-to-date item-item similarity matrix Sim, which is the core of online recommendation. As shown in Fig. 1, the ETL-Recommendation pipeline is done in 4 steps (i.e., ETL, Pref, CF, Test).

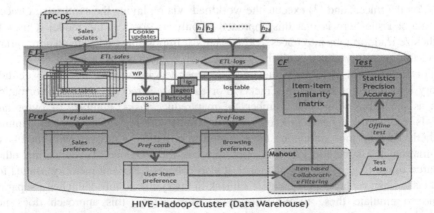

Fig. 1. Architecture of ETL-Recommendation

1. **Data Preparation** generates both the base and update data for sales, logs, and cookies on HDFS. The base and update sales data are first generated by distributed running of DSGen in TPC-DS [9]. Then the base and update logs data is generated by 100x scaling of sales records, indicating that 100 times of Web page browsing result in 1 online product buying. Note that the logs update data is partitioned into 24 HDFS files to simulate the hourly collecting of Web logs data from Web servers. The cookies data maintains the mapping from cookieid to customerid. Once all base data and update data generated in HDFS, the base data file will be put into Hive warehouse as base tables. The prepare process guarantees the data consistence within daily update data.

2. **ETL** simulates the daily workloads to refresh the structured sales updates (by *ETL-sales*) and unstructured logs updates (by *ETL-logs*) to warehouse. In particular, the *ETL-sales* step collects the buying transaction from structured database, while the *ETL-logs* step captures one day's user browsing history segmented by 24 h and append the result into logs fact table as a new partition in warehouse. In current stage, we only involve the Web sales data but not the store/catalog sales data. Note that *ETL-logs* holds until *ETL-sales* ends as they are data dependence on the Web_pages table. Since Hive does not naturally support the record update/delete operations on Hive tables, we wrote additional Hive queries to implement the functionality of record update/delete operations. The uniformed apache log format looks like

```
10.6.38.8 - - BAF42487E0C7076CE576FAAB0E1852EC [14/Dec/1999 00:00:00
-0] "GET ?item=8745 HTTP/1.1" 101 1345 http://www.foo.com?item=8745
"Mozilla/4.0 (compatible; MSIE 5.5; Windows 98; Win 9x 4.90)"
```

3. **Pref** extracts the user buying and browsing histories, and combines them to generate the user-item preference table in warehouse. The weighted combining of user-item preference is based on the assumption that buying behavior (with quantity) implies 10x user-item preference while Web browsing of product implies only 1x user-item preference. The customerid is determined by JOINing table *logs* with *cookies* on key cookieid. *Pref-logs* can be actually incremental since *logs* table is partitioned on DATE.

4. **CF** step simply utilizes the item-based collaborative filtering algorithm in Mahout to generate the item-item similarity matrix (*Sim*) (Fig. 2).

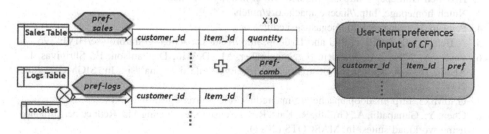

Fig. 2. User preference calculation

5. *Test* does the offline validation on *Sim* with the present of pre-defined test data. The test data basically includes the user status vectors and user clicks history generated by data preparation. However, note that the regression in test result does not mean real regression in realtime (online) recommendation, while the improvement in test result does not mean promising improvement in realtime recommendation.

In current implementation, the pipeline repeats 7 times to simulate the ETL-Recommendation workloads on warehouse in a period of a week. At the end of pipeline running, an evaluation report will be generated with the information of base data size, refresh data size, time cost, average throughput, and corresponding time cost of each pipeline stage.

For the ETL-Recommendation pipeline, only the structured input data are generated by TPC DSGen, while the unstructured log data are generated by our own implemented programs and they are in much larger size (i.e., logically 100x of structured data). The recommendation related components in the pipeline are not defined or included in the TPC-DS benchmark. The pipeline can be viewed as an end-to-end Hadoop application to evaluate the capability of Hadoop-based big data framework and underneath system.

3   Future Work

In future, we would like to explore several improvements to HiBench based on the user feedbacks, including broader coverage of Hadoop applications, evaluations of the interactions of concurrent Hadoop jobs. We also consider including other commonly used offline recommendation algorithms, and integrating the online system for real-world recommendation engine evaluation. Furthermore, we would also improve the metrics system to produce more detailed and meaningful evaluation result for customers.

References

1. Huang, S., Huang, J., Dai, J., Xie, T., Huang, B.: The HiBench benchmark suite: characterization of the MapReduce-based data analysis. In: ICDEW, Hibench, March 2010
2. HiBench Homepage. https://github.com/intel-hadoop/HiBench
3. Nutch homepage. http://lucene.apache.org/nutch/
4. Pegasus Homepage. http://pegasus.isi.edu/
5. A Benchmark for Hive, PIG and Hadoop. http://issues.apache.org/jira/browse/HIVE-396
6. Pavlo, A., Rasin, A., Madden, S., Stonebraker, M., DeWitt, D., Paulson, E., Shrinivas, L., Abadi, D.J.: A comparison of approaches to large-scale data analysis. In: SIGMOD, June 2009
7. GridMix3. http://hadoop.apache.org/mapreduce/docs/current/gridmix.html
8. Chen, Y., Ganapathi, A., Griffith, R., Katz. R.: The case for evaluating MapReduce performance using workload suites. In: MASCOTS (2011)
9. TPC Benchmark DS (2012)

Big Data Benchmark - Big DS

Jun-Ming Zhao[✉], Wen-Shuan Wang, Xian Liu, and You-Fu Chen

HP Building, No 112, Jianguo Road Chaoyang District, Beijing 100022, China
{jimmy.zhao,wenshuan.wang,shine.liu,
youfu.chen}@hp.com

Abstract. Performance and scalability in clusters of heterogeneous and complex Big Data Analytic environments are always unpredictable. In this paper, we are trying to address this problem by using a benchmark named "Big DS". The benchmark adopts many great ideas from some famous industry benchmarks like TPC-H [1], TPC-DS [1], SPECvirt_sc2010 [2] and SPECjbb2005 [2], we also adopt some ideas from non-standard benchmarks liked TeraSort [3], SWIM [4], etc. By defining a configurable workload for different big data analytics environment, Big DS can be used for measuring the performance and scalability of a big data analytics platform or environment for different business.

Keywords: Big data · Analytics platform · Analytics engine · Benchmark · Performance · Scalability · Analytics · TPC [1] · SPEC [2] · TPC-DS [1] · Terasort · SWIM [4] · Hadoop [5] · Hive [6] · Impala [7] · Data warehouse · BI

1 Introduction

Since Google published BigTable [8], a lot of new Big Data technologies and solutions have been innovated from different communities or vendors, with different programming languages and different features. The most famous one of them is Hadoop [5]. You probably can see a new name of Big Table, NoSQL or NewSQL jump into your eyes every 3 months, claiming different brilliant features. But it is hard to tell which technology or solution has better performance and scalability.

At the same time, there are many requirements for a meaningful benchmark. Karl Huppler [9] defined the following set of attributes of a good benchmark:

- Relevant
- Repeatable
- Fair and portable
- Economical
- Verifiable

While designing Big DS, the above requirements were taken into consideration. Specifically they were implemented as follows:

- Relevant

    - Real world environments – The target environment of Big DS is a hybrid and scale-out data mining system/cluster. Just like most of big data environments,

© Springer International Publishing Switzerland 2014
T. Rabl et al. (Eds.): WBDB 2013, LNCS 8585, pp. 49–57, 2014.
DOI: 10.1007/978-3-319-10596-3_5

the data will be stored in a cluster and it will also be analyzed in the same system.

- Real world workloads - The benchmark need to be a good sample of the real world workloads. In the big data area, there are so many different use cases. The workload should be able to handle huge volume data, the data will be integrated using the similar way of real world; the workload should be able to handle different data structure with different size.
- Performance and Scalability of the whole cluster – Performance and scalability of target systems are essential for cluster-oriented benchmarks. Benchmark needs to show different capability of target system and their scalability for better predicting future usage of the system.

- Repeatable and verifiable

  - Repeatability and verifiability of benchmark results are always important, no matter what type of workload is measured. In Big DS, we will reuse the successfully used business model from TPC-DS with important updates, which simulate the social and mobile impact to the business model of the real world. The micro-benchmarks will be re-designed carefully, so that their test result can keep consistent in different runs in the same configuration.

- Economical

  - Big data clusters tend to be expensive, even though each node of the cluster might be cheap. So, while designing the benchmark, we are trying to use a thin driver design to reduce the cost of the overall system. And, the benchmark will include performance per dollar and performance per watt metrics, so that benchmark publishers will use more cost conscious system configurations. To be more specific, total cost of the system and total power consumption of the cluster will be included in the benchmark report.

2 Problem Statements

2.1 Gaps Analysis

Here we list the typical problems or requirements for a benchmark while choosing a big data analytics system, from different perspectives of big data customers. We also list the solutions to fill the gaps:

- Buyer of Big Data Analytics System:

  - Problem 1: Complicated and different business models, configurations, execution processes and metrics. It's really hard to compare with each other.
  - Solution: We need a well-defined and trustable business model; a proved fix process to execute; a vendor independent design; and fully disclosed and auditable result.

- Big Data Solution Vendors

  - Problem: How to easily show my own value?
  - Solution: Common metrics and sub metrics. Well-defined specification and guidance but allow customization for different technologies. Fair, repeatable and verifiable.

- Standard Organizations

  - Problem: There's no one fit-for-all solutions for different customers.
  - Solution: Flexible framework and proven-models of the new environment and trustable benchmark(s).

3 Big Data Benchmark - Big DS

3.1 The New Generation of Data Analytics Process

The evaluation of technology brings us the so called Software Defined Networt/Data Center, and infrasctructure of big data systems can defintely be considered as the most important one of them. Open source solutions liked Hadoop or traditionaly BI techlogies are merging their strengths together. Traditional BI vendors liked Teradata or HP Vertica starting to embed Hadoop as their data integration component or as a plugin. We can see more Hadoop or similar technologies vendors support better analytics features in their solutions. Cloudera Impala [7] or Apache Drill [13] are the top players in this area.

Figure 1 below shows one of the changes to the existing BI process defined based on what we observed from Hadoop, Spark and other systems. And we think this change will be a common case for those big data technology vendors.

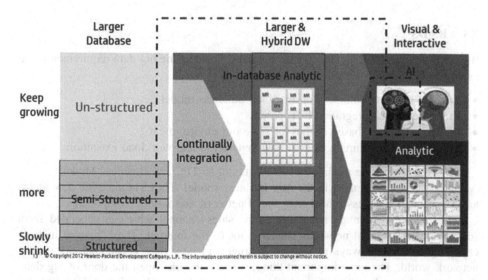

Fig. 1. Model of Modern Hybrid Big Data Platform (The dashed box includes the changes we consider for a big data benchmark.)

Figure 1 shows a model of big data analytics system. The whole environment is a hybrid big data system, where data is stored and analyzed in the same cluster.

On the left are the types of data that will be loaded, extracted, transformed and analyzed in the environment, including unstructured data, semi-structured data and structured data. To better simulate the real world big data analytics environment, the data should be more LIVE data or real time data. New data will be integrated continuously. And visualization tools or other machine learning tools will be used to directly analyze the live data and generate report or tabular data from the system.

On the right side, there is a smaller dashed box, which represents the machine learning will also be supported and executed in the analytics platform.

This is the environment to meet modern big data analytics requirements.

3.2 Starting from TPC-DS

TPC-DS is an evolution edition of TPC-H benchmark. Its predecessor is adopted widely in the database world. The benchmark resembles a decision support system. It covers basic procedures of a generic BI process.

TPC-DS is an existing industry standard benchmark, which is designed for bigger and more complex BI environments. The business model is a good sample of the real world workload. And the data scaling and workloads also match the real world BI process pretty well. From the big data perspective point of view, TPC-DS design meets the 3Vs (volume, velocity and variety) of Big Data. Founding Big DS on TPC-DS can meet most of the modern cooperation's Big Data requirements. This will also increase its acceptance in the industry as a big data benchmark.

The next section will describe how we extend TPC-DS to BigDS, and why we extend them.

3.3 BigDS

BigDS is an extension of TPC-DS for big data. To meet the big data requirements, the extension includes:

- Increase the data volume with modified business model
- Brand new data integration/ETL process
- Cloud benchmark based design for scale out environment
- Configurable benchmark in both data generation and workload execution

Figure 2 shows the modified business model to TPC-DS Business Model:

There are four domains in the new business model. They simulate the new architecture of retail business model with social network and mobile components.

The marketing domain simulates a new sales channel which is generated from social networks. Social networks change a lot to the internet. They changed and are changing customers' ways to use the Internet and how they spend their life in the network world. The most important thing, social networks open the door of big data. Customers start to log their life in the Internet. They share, talk and show huge amount

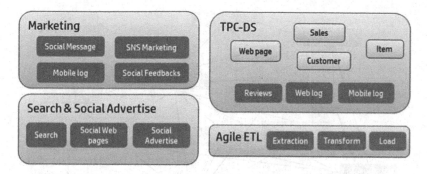

Fig. 2. Big DS Business Model Diagram

of information of themselves and the world around them in social network. This heavily changes the way how we understand our customers. So this is the most important change we make to the original business model. We keep this part as a new channel and it's different with the original web sales channel.

The second new domain is called Search and Social Advertisement domain. In this domain, we try to simulate the way that how we bring our products to our customer. To avoid conflict with the old web sales channel, it will be more focused on the advertisement of social network websites. Company with the social components, we added in the new social marketing and search components. They will simulate different user behaviors and actions which can drive to a final purchase.

The TPC-DS domain is almost the same as the old business model. To be able to know more about our customer and internal business efficiency, we add a few components to generate more data from existing components. The dark boxes are new added component in the diagram. They also represent for different kind of data structure, including unstructured data and semi-structured data.

The fourth domain is what we called Agile ETL part. This part will simulate how the data will be prepared and processed before entering the real analytics process. We will explain more in below section.

3.4   Modified Business Model

In upcoming years more and more companies will put part of their marketing activities on those social networking websites and mobile apps. Huge social networking websites like Facebook or Tencent will produce a huge amount of click stream data on their website, and this also brings a new way for customer to buy things. The original retail system design of TPC-DS has a well-defined data structure. We can easily extend the TPC-DS data schema with social networking and mobile parts. By doing this change, the data schema of BigDS will be largely increased, creating large opportunities for modeling a complex but more realastic big data analytics environment.

The data structure of social networking and mobile data are composite of structured data, unstructured data and semi-structured data, but a large portion of them are semi-structure data and unstructured data.

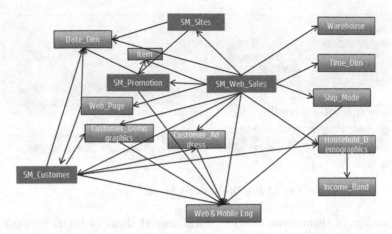

Fig. 3. Relationship of data model with social and mobile data

Figure 3 shows the changes we made to the old table model. The dark boxes are new tables.

They are the base tables of the new social sales channel, all other social and mobile tables will be added based on them. With these tables, we will be able to add more social customer behavior data in the future based on the micro-benchmarks we want to simulate.

The new added tables will also introduce some changes to existing tables, but those are all minor change. It won't change the original behaviors and business transactions of the TPC-DS business model.

3.5 Agile ETL – New Way of Data Integration

As we mentioned, modern big data analytics systems are more focused on live data analytics. So in this benchmark, the data integration process will be largely different compared to what TPC-DS uses. What we call Agile ETL is the way that we inject data into the analytics platform will have finer granularity and will be executed more frequently.

In the real world of big data analytics, analysts will observe and analyze the business, then decide to collect new data, and then develop new analysis component based on those data. New data collectors will be developed in hours, and the report or dashboard might also be developed in hours, or even shorter time. Analytics platform needs to handle the new data and start to analyze them quickly and easily. This is really different from the traditional BI process. A huge amount of data will be collected and loaded quickly. Some of the data extraction and transforming might be done in the same data warehouse in the big data cluster. The efficiency and effect of the new analysis will be known in a short time.

To simulate agile ETL processes well, we will categorize the data generator into different categories. Each category might be executed in different frequency so that they can better simulate the real world workloads:

- Data generator will be categorized based on the final purpose of report. Reports will be defined as real time report, hourly report, daily report, weekly report and monthly report. Different report will be executed in different frequency.
- Data generator will be also categorized based on their data size and data structure. Bigger data might cause more system resource consumption. In real environment, they will be executed at night. In our benchmark, there's also a "night" time designed.

As what TPC-DS does, Agile ETL component will be also considered as part of the workload, and its performance will be counted into the final score of performance metrics.

3.6   Other Considerations

Scalability. In most cases, big data analytics platform are scale-out environments, besides knowing the peak performance of the environment, knowing how well the environment can scale is also an important metric for benchmark user. So BigDS will have multiple metrics, scalability metrics will be one of them.

While designing the scalability metrics, we want to leverage other good benchmark examples. There are many other great benchmarks already available on the world that has many great characters. SPECpower_2008, SPECjbb2005, SPECjbb2013 and SPECvirt_sc2010 are good examples of them. In a big data analytics platform, we think these two metrics can show the scalability of an environment:

- How big the data platform can handle?
- How many jobs the platform can handle?

As a big data benchmark, in most cases, our environment is big can might be easily saturated with small workload. By adopting the idea of benchmarks we mentioned above liked SPECvirt_sc2010, we will use similar idea in BigDS.

We will use a concept called "Copy" in BigDS. While designing BigDS, there is a base micro-benchmark set (the green box in Fig. 4) and a base set of data generators (the blue box in Fig. 4). To saturate the cluster, the base set of data generator and workload can be executed with multiple copies at the same time. How many copies can be executed in the cluster will be the scalability metrics of BigDS. While adding data generators and micro-benchmarks, all of them will not impact each other and can be executed in parallel.

The base copy of micro-benchmarks is composed with three different kind of workloads:

- Ported TPC-DS queries
- SQL liked queries, i.e. Hive [6] Queries
- Analytics jobs, i.e. Map Reduce jobs or other machine learning jobs.

Figure 4 diagram shows one of ways of BigDS execution. In that way, we will generate data first, and then executed the base set of micro-benchmark. To know the

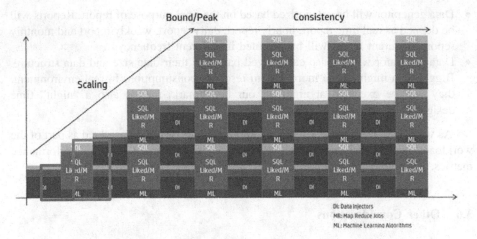

Fig. 4. Workload Scalability Test Pattern (Color figure online)

scalability, we will then increase the copy number of workloads step by step until the copy number we defined.

There's another way that we will execute the workload – by the frequency of the report. The workload will be executed using the similar way as below but more phases. The exactly phases will be determined after all micro-benchmarks are characterized.

Configurable Benchmark. Customer data might come from different sources. We designed different structured of data in BigDS, and will develop different kind of micro-benchmarks. But those micro-benchmarks might not be fixed for all BigDS users. So in our future design, user will be able to enable or disable a certain micro-benchmark if you only tend to use BigDS as a performance tool to know performance of your environment. If users want to compare the result with others, then it needs to run fully list of all defined micro-benchmarks with required execution scenario.

4   Summary

BigDS is an experiment for evaluating the performance and scalability of big data platform. We define some characteristics of big data analytics platforms so that we can have a more fixed and common configuration from thousands of big data technologies and vendors.

The design leverages some great ideas from the existing benchmark TPC-DS [1], BigBench [11], SPECvirt_2010sc [2], SPECjbb2005 [3], etc., adding many attributes of real big data environment which are essential for most of modern big data business environment. In this paper, we define the data model, workload characters and workload execution process of typical big data analytics platform.

The benchmark is still under development. Ongoing work includes adding more micro-benchmarks and corresponding data generator, and pick up a best way which can really measure the performance and scalability of a big data analytics platform.

References

1. TPC. TPC is a trademark of the Transaction Processing Performance Council. TPC-H and TPC-DS are the decision support benchmarks of TPC organization. http://www.tpc.org
2. SPEC. SPEC is a trademark of the Standard Performance Evaluation Corporation 1995–2014. SPECjbb2005 is the server side Java Benchmark of SPEC.org. SPECjbb2013 is the evaluation version of SPECjbb2005. SPECvirt_2010sc is the server consolidation virtualization benchmark of SPEC.org. http://www.spec.org
3. TeraSort. Refer to the Apache Terasort benchmark, which is a MapReduce version of Sort benchmark
4. SWIM. SWIM stands for Statistical Workload Injector for MapReduce. The synthesis methodology is adopted in BigDS and it's supporting toolset
5. Apache Hadoop and it's related projects. Apache Hadoop is an open-source software framework for storage and large-scale processing of data-sets on clusters of commodity hardware. Hadoop is an Apache top-level project being built and used by a global community of contributors and users. It is licensed under the Apache License 2.0
6. Apache Hive. Apache Hive is a data warehouse infrastructure built on top of Hadoop for providing data summarization, query, and analysis. [1] While initially developed by Facebook
7. Cloudera Impala. loudera Impala is Cloudera's open source massively parallel processing (MPP) SQL query engine for data stored in a computer cluster running Apache Hadoop. http://www.cloudera.com/content/cloudera/en/products-and-services/cdh/impala.html
8. Google BigTable. Refer to Google's BigTable paper. http://research.google.com/archive/bigtable-osdi06.pdf
9. Huppler, K.: Chairman TPC. The author of "The art of building a good benchmark" (2009). http://www.tpc.org/tpctc/tpctc2009/tpctc2009-03.pdf
10. WBDB, Workshop of Big Data Benchmarking, San Jose. http://clds.ucsd.edu/wbdb2012
11. Big Bench. Extend TPC-DS specification to include unstructured and semi-structured data; modify the TPC-DS. In: A data model for BigBench was proposed in the First WBDB Workshop by Ghazal (2012)
12. Deep Analytic Pipeline. A Benchmark Proposal by Milind Bhandarkar (Pivotal Chief Scientist), (2013). http://clds.sdsc.edu/sites/clds.sdsc.edu/files/2013-03-07-DeepAnalytics Pipeline.pdf
13. Apache Drill Project. Apache Drill is an open-source software framework that supports data-intensive distributed applications for interactive analysis of large-scale datasets. Drill is the open source version of Google's Dremel system which is available as an IaaS service called Google BigQuery. http://incubator.apache.org/drill/
14. Google Dremel. http://research.google.com/pubs/pub36632.html
15. Google Big Query. https://developers.google.com/bigquery/

WGB: Towards a Universal Graph Benchmark

Khaled Ammar[✉] and M. Tamer Özsu

Cheriton School of Computer Science, University of Waterloo,
Waterloo, ON, Canada
{khaled.ammar,tamer.ozsu}@uwaterloo.com

Abstract. Graph data are of growing importance in many recent applications. There are many systems proposed in the last decade for graph processing and analysis. Unfortunately, with the exception of RDF stores, every system uses different datasets and queries to assess its scalability and efficiency. This makes it challenging (and sometimes impossible) to conduct a meaningful comparison. Our aim is to close this gap by introducing Waterloo Graph Benchmark (WGB), a benchmark for graph processing systems that offers an efficient generator that creates dynamic graphs with properties similar to real-life ones. WGB includes the basic graph queries which are used for building graph applications.

1 Introduction

Graph data are of growing importance in many recent applications including semantic web (i.e., RDF), bioinformatics, software engineering, e-commerce, finance, social networks, and physical networks. Graphs naturally model complicated structures such as chemical compounds, protein interaction networks, program dependence, product recommendation, fraud detection, social networks, computer networks, and web page connections. The size and complexity of these graphs raise significant data management and data analysis challenges. There has been a considerable amount of research in the last decade in analyzing and processing graph structures in real applications which leads to the development of a number of alternative algorithms, techniques and systems.

The validation of these algorithms/techniques/systems, and their comparison with each other, requires the use of a standard graph data set and a standard set of queries – i.e., a benchmark. Unfortunately, with the exception of RDF stores, every graph analysis system that has been proposed uses different datasets and queries to assess its scalability and efficiency. This makes it challenging (and sometimes impossible) to conduct meaningful comparisons.

In this paper, we propose a universal graph benchmark, called Waterloo Graph Benchmark (WGB), that is suitable for testing (parallel) graph systems, and that includes a wide range of operational, traversal and mining queries. A benchmark consists of two parts: (a) a data generator, and (b) a workload specification. The data generator produces the actual data over which the queries in the workload are executed for performance evaluation. It is generally desirable for the data and the workload to represent real applications. Furthermore,

© Springer International Publishing Switzerland 2014
T. Rabl et al. (Eds.): WBDB 2013, LNCS 8585, pp. 58–72, 2014.
DOI: 10.1007/978-3-319-10596-3_6

a benchmark should be complete, efficient, and usable. WGB features these requirements. It includes a representative query for each class of queries that might be used in graph data processing. We define a simple schema in Sect. 3 and specify multiple queries based on it. These are fundamental and common graph queries that many graph systems already support. WGB can efficiently generate very large datasets using limited resources in a short time, and can be easily tested by various graph systems with simple and straightforward interfaces. Although various data sets have been available for testing graph data management systems, WGB includes a simple and efficient data generator. It is the only graph generator that creates time-evolving graphs, that respects the power-law distribution for vertex degrees, and that provides full freedom in terms of nodes types, structure level, density and graph size.

In the following section, we discuss the existing benchmarks and clarify the motivation behind proposing a new one. Section 3 discusses the three main types of queries in graph systems and presents queries included in WGB that represent each class. Section 4 includes a discussion of the data generator followed by experiments to validate it. Section 6 briefly discusses the existing parallel/distributed graph systems, and Sect. 7 summarizes the paper and our future plans for WGB.

2   Existing Graph Benchmarks

We briefly review in this section the proposed benchmarks for graph analysis, pointing out some of their issues that guide us in the development of WGB. There are also a number of real-life datasets available in the literature such as those published in the Stanford Large Network Dataset Collection[1]. These real data sets are complementary to a benchmark. In general, one can use any available real-life dataset with a benchmark, but these datasets cannot be modified to meet different needs.

2.1   HPC Scalable Graph Analysis Benchmark

The idea behind HPC Scalable Graph Analysis Benchmark[2] is to develop an application with multiple kernels that access a weighted directed graph. These kernels represent multiple graph queries. They require irregular access to the graph so it is not expected that a single data structure for the input graph dataset would be optimal for all kernels. There are four kernels designed to address the following operations: bulk insertion, retrieving highest weight edges, k-hops operations, and betweenness calculation. However, the benchmark does not include many other graph operations such as updates or iterative queries [1].

---

[1] http://snap.stanford.edu/data/
[2] www.graphanalysis.org/index.html

2.2 Graph Traversal Benchmark

The authors of the Graph Traversal Benchmark [2] classify graph data management systems in two categories: (1) graph databases such as OrientDB[3], DEX[4], and Neo4j[5], and (2) distributed graph processing frameworks such as Pregel [3] and GraphLab [4]. This benchmark focuses on traversal operations on graph databases. Consequently, the capability of this benchmark is limited, because it does not support graph data queries or graph analysis algorithms. Moreover, it does not consider distributed graph processing frameworks.

2.3 Graph500

Graph500[6] is developed by a group of experts from academia, industry, and research laboratories. The main target is evaluating supercomputing systems for data intensive applications, focusing on graph analysis, to guide the design of hardware architectures and their software systems. Currently there is only one benchmark included in Graph 500, which is the HPC Scalable Graph Analysis Benchmark discussed earlier.

2.4 BigBench

BigBench [5] is a new benchmark that includes structured, semi-structured and unstructured data. The structured schema is borrowed from the Transaction Processing Performance Council's Decision Support benchmark (TPC-DS), the semi-structured part has user clicks on the retailer's website while the unstructured data has the online product reviews. The benchmark also includes a data generator and business queries for its schema. Although this benchmark covers a wide range of queries and data types, these are not graph-structured hence cannot be used as a graph benchmark.

2.5 BigDataBench

BigDataBench [6] is a big data benchmark suite. It is generic and includes various types of dataset: text, graph, table. The graph component uses Kronecker graph model [7] to create synthetic datasets based on real datasets. In this benchmark, the data generator is developed to learn the properties of a real graph and then generate synthetic data with similar characteristics [8]. Two real graph datasets are included: Google web graph, and Facebook social graph. Although Kronecker graph generator is able to create graphs similar to real ones and has a very good mathematical model, it has some disadvantages such as generation of multinomial/lognormal degree distributions instead of power-law

---

[3] http://www.orientdb.org/

[4] http://www.sparsity-technologies.com/dex

[5] http://www.neo4j.org

[6] http://www.graph500.org/ (accessed on 24th May 2013)

and being not flexible [9]. WGB generator, on the other hand, creates power-law distributions and is very flexible. It is capable of generating directed/undirected, weighted/unweighted, and bipartite/unipartite/multipartite graphs. Our benchmark is a domain specific big data benchmark focusing on graph data only. We need a flexible generator because we aim to assess graph systems using all potential graph datasets that may comply with known real graph properties.

2.6 RDF Benchmarks

For RDF stores, a number benchmarks have been proposed, such as the Berlin SPARQL Benchmark [10], DBpedia SPARQL Benchmark [11], LUBM [12], and SP^{2}Bench SPARQL [13]. These exclusively focus on RDF graphs whose properties are different than graphs seen in other application domains. It is interesting to note that the properties of RDF graphs used in these benchmarks are also different than the properties of real RDF graphs [14]. This has resulted in development of more realistic RDF benchmarks, such as WSDTS [15]. Furthermore, there are differences in the workloads. SPARQL is the standard query language for RDF systems and all benchmarks focus on SPARQL queries. These queries usually have specific structures (e.g., star shape) that are not evident in more general graph queries. Moreover, SPARQL cannot represent some popular graph algorithms such Pagerank.

3   Workload Characterization

There are three main workload categories in graph applications: online queries, updates, and iterative queries. *Online queries* are read only, interactive, usually do not have high complexity, and do not require multiple computation iterations – the answer should be computed quickly. *Updates* are also interactive, but they change the graph's structure by changing its parameters or its topology. Updates also include adding or removing a vertex or an edge. *Iterative queries* are heavy analytical queries, possibly requiring multiple passes over the data, and usually take a long time to complete. They may include some updates but the changes to the graph does not need to occur in real time.

   In this section, we describe these queries assuming a relational database schema *Graph* that has the following three tables:

- *Nodes{NodeId, {p_i}}*, holds all the nodes in the graph and the properties of each node {p_i}.
- *Edges{EdgeId, from, to, {p_j}}*, holds the edges. For each edge, there are two foreign keys to the Nodes table pointing to the source (*from*) and destination (*to*) of this edge. Similar to nodes, edges may have a list of properties {p_j}.
- *ShortestPath{from, to, length, path}*. This table maintains the length and sequence of the shortest path between any two nodes in the graph. If the two nodes are not connected, i.e., not reachable, then the values of columns *length* and *path* are *null*.

It is important to note that we use these tables only to formalize the queries. The actual implementation of the graph data structure is up to each system as long as the result of each query is correct with respect to this schema.

Since there is no single programming language that all graph systems use (analogous to SPARQL for RDF), it is not possible to specify the queries in a language. Thus we specify queries abstractly using SQL over the schema specified above.

3.1 Online Queries

Most graph applications include only online queries such as a simple check of the availability of customers' relationship to a retail shop, or an update for the news feed in social media, etc. This class of queries also include graph matching queries.

Find Matching Nodes (Edges). The objective of this query is to find specific node(s) or edge(s). This is an online query and can be described in an SQL-like notation as follows. Note that X is a list of given parameter values.

```
SELECT NodeId FROM Nodes WHERE { p[i] = X[i] }
```

```
SELECT EdgeId FROM Edges WHERE { p[j] = X[j] }
```

The complexity of this class of queries vary based on the data structure(s) used in a system.

Find k-hop Neighbours. This query looks for neighbours that can be reached by up to k-hops from a specific node. It could be merged with the previous query to find neighbours who satisfy some conditions. A k-hop query from a given node u is the following:

```
SELECT S.to
FROM ShortestPath S
WHERE  S.from = u  AND S.length < (k+1)
```

Reachability and Shortest Path Query. For any two nodes $u, v \in G = \{V, E\}$, the reachability query from u to v returns true if and only if there is a directed path between them. A reachability query can be specified as:

```
SELECT true
FROM ShortestPath S
WHERE S.from = u AND S.to = v
AND S.length is not null
```

For any two nodes $u, v \in G = \{V, E\}$, the shortest path query returns the shortest path between u and v. The shortest path query between two nodes u and v can be written in SQL as:

```
SELECT path
FROM ShortestPath S
WHERE S.from = u AND S.to = v
AND S.length is not Null
```

There are two straightforward approaches to answer these two queries:

1. Build a breadth- or depth-first search over the graph from node u. This approach costs $O(|V| + |E|)$ to answer a query, where $|V|$ is number of nodes and $|E|$ is number of edges.
2. Build an edge transitive query closure of the graph to answer reachability queries in $O(1)$ time complexity which requires space $O(|V|^2)$. The transitive closure works as an index.

Most of the existing work on this problem try to optimize between these two extremes. There are many proposed approaches to reduce the index size and to minimize the complexity of answering the query [16].

In our experiments, we implemented these queries using breadth-first search, because at this stage we would like to eliminate the benefit of indexes and evaluate simply how a graph system performs traversal. In future, we may extend this to evaluating queries based on possible indexes in the system.

Pattern Matching Query. Given a data graph G and a query graph Q with n vertices (each query can be represented as a graph), any set of n vertices in G is said to match Q, if: (1) all $n \in G$ have the same labels as the corresponding vertices in Q; and (2) for any two adjacent vertices in Q, there is a path between the corresponding vertices in G. This is sometimes referred to as the *reachability pattern matching query*. A variation, that is called *distance pattern matching query*, introduces an additional query parameter δ such that for any two adjacent vertices in Q, the shortest distance between the corresponding vertices in G should be less than or equal to δ. A reachability pattern matching query input and output can be described as follows:

Input: Q specified as a list of edges $(FromLabel_i, ToLabel_i)$
Output: List of matching sub-graphs, each of which consists of a list of edges $(from_i, to_i)$ | Label$(from_i) = FromLabel_i$ and Label$(to_i) = ToLabel_i$ and Reachability$(from_i, to_i) = true$.

In the above description we use the Reachability(u, v) to represent a reachability SQL query from u to v. Similarly, we use Label(id) as a shorthand for the following SQL query:

```
SELECT label
FROM Nodes
WHERE NodeId = id
```

3.2 Updates

This category is simple but very important to evaluate a system's readiness for handling time-evolving (dynamic) graphs. Most available graph systems do not support online updates because these queries may invalidate indexing and partitioning decisions. A notable exception is G* [17]. A number of different updates are possible:

- Adding a new edge requires checking if either of its two nodes need to be added.
- The complexity of adding a new node depends on the underlying data structure. Many implementations use adjacency-vertex list. In this case, adding a new node simply means adding a new item to this list.
- Updates to physical data structures. Some systems store indexes to efficiently handle queries such as shortest path and reachability, these indexes should be maintained when a node (edge) is added to the graph.
- Updating and removing a node (edge) requires finding this node (edge) first then updating its information. Similar to adding a node (edge), this will lead to updates for all maintained indexes.

The set of updates can be specified as follows, where X is a list of given values for a node's or edge's parameters, u is a given node, and e is a given edge.

- `INSERT INTO Nodes VALUES (NodeId, { p[i] } )`
- `INSERT INTO Edges VALUES (EdgeId, from, to, { p[j] } )`
- `UPDATE Nodes SET { p[i] = X[i] } WHERE NodeId = u`
- `UPDATE Edges set { p[j] = X[j] } WHERE EdgeId = e`
- `DELETE FROM Nodes WHERE NodeId = u`
- `DELETE FROM Edges WHERE EdgeId = e`

3.3 Iterative Queries

Iterative queries are usually analytical workload over graph datasets (graph analytics). There is a large number of analysis algorithms such as subgraph discovery, finding important nodes, attack detection, diameter computation, topic detection, and triangle counting. There are several alternatives for each of these problems. In our benchmark we include two iterative graph analysis algorithms that represent the main theme of graph mining algorithms, *"iterate on graph data until convergence"*. These two algorithms are Pagerank and Clustering.

Pagerank. Pagerank is the prototypical algorithm to check a graph system's performance in performing iterative queries. Almost all proposed graph analysis systems use it to evaluate their performance. In a nutshell, Pagerank finds the most important and influential vertices in a graph by repeating a simple mathematical equation for each node until the ranks of every node converge to a fixed-point.

Clustering. There are two main types of clustering algorithms: *node clustering* ones and *graph clustering* ones [18]. Node clustering algorithms operate on a single graph that they partition into k clusters such that each cluster has similar nodes. Graph clustering algorithms operate on (possibly large) number of graphs and cluster them based on their structure.

Given a connected graph, the main task of a clustering algorithm is to put similar graph nodes in the same group/cluster by minimizing an optimality criterion such as number of edges between clusters. Partitioning a graph into k clusters where $k \leq 2$ is solvable, as it can be mapped to the minimum cut problem. However, it is significantly more difficult, and is NP-hard, to optimally partition a graph when $k > 2$. Local optimum can be achieved by following techniques similar to k-medoid or k-means algorithms [18].

At this stage, we do not consider graph clustering algorithms in WGB. We have two clustering algorithms: the *Connected components* algorithm and the classical *k-means* algorithm. Connected component algorithm is a popular straightforward approach in node clustering where any two nodes belong to a cluster if there is an edge connects them. K-means algorithm picks k random data points as seeds and then assigns every other data point to one of these seeds (clusters). Guided by an objective function, it keeps searching for better seeds and move points between clusters. Its objective function is to minimize the sum of the distances between data points in a cluster [18].

4   Data Generation

We propose a new graph generation tool that creates graphs with real-life properties and allows users to control their behaviour. Our graph generator is based on RTG [19], which expands Millar's work [20]. Millar showed that typing random characters on a keyboard with only one character identified as a separator between words can generate a list of words that follow a power-law distribution. RTG builds on this by connecting each pair of consecutive words by an edge. The final result is a graph whose node degrees follow the power-law distribution. RTG enhances the properties of the generated graphs by introducing some dynamic graph properties such as shrinking diameter and densification [21]. Graph diameter is the longest shortest path between any two vertices in the graph. Shrinking diameter indicates that a graph's diameter decreases as the graph grows over time. Graph density is the ratio between the number of existing edges in a graph and the maximum possible number of edges assuming no multiple edges between any two vertices in the graph. Densification means that a graph's density increases over time.

Although the RTG generator is flexible and can generate graphs with real-life properties, it inherits several required parameters from Miller's process which are not strongly associated with graphs. Based on Millar's work, the generation process needs two main attributes: number of characters (k) and occurrence probability of the separator character (q). Setting these parameters to match desired graphs had not been considered. Our generator addresses this issue by

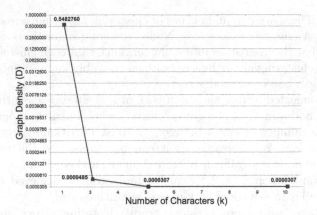

Fig. 1. The relation between the number of characters (k) and the density of the generated graph

only considering parameters directly related to graph properties, and then using approximation models to identify the other ones.

Experimentally we found that $k \in [5, 10]$ is a reasonable choice for generating many massive graphs. Increasing k too much does not have any significant influence on the generated graphs. A very small k generates graphs with high density (see Fig. 1), which is not realistic, and also influences their power-law distribution. A similar result was proven by Millar, where it was shown that changing k from 3 to infinity does not have a significant impact on the generated power-law distribution.

In the following we discuss the relationship between the graph density D and the occurrence probability of the separator character q. The following two equations were already proven [19]. W is the total number of generated edges in the graph, p is the occurrence probability of each character (except the separator character) and is a function of q and k such that $1 = q + k \times p$, N is number of unique vertices and E is number of unique edges:

$$N \propto W^{-log_p k} \tag{1}$$

$$E \approx W^{-log_p k} \times (1 + c' \times log\ W), c' = \frac{q^{-log_p k}}{-log\ p} > 0 \tag{2}$$

Graph density, D, is defined as $D = \frac{E}{N \times N - 1}$ assuming that self-loop edges are not allowed. Based on the above equations, we can write $D = fn(q, k, W)$. Then, we can show that $q = fn(D, k, W)$. However, these functions are very complex and already approximate. Taking the inverse of the D-function to get the q-function is difficult if not impossible. Therefore, we use an approximation technique based on Gaussian models. Based on a large training dataset, we built a Gaussian model to predict q based on D and W. In our experiments, the average error in predicting q was only 10 %.

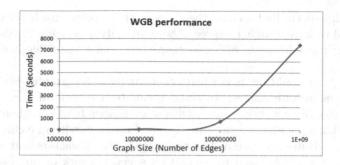

Fig. 2. Generator performance with respect to number of generated edges

Our generator has seven parameters: graph weight as total number of generated edges (W), partite flag (PF) which identifies if the graph should be unipartite or multi-partite, number of node types (NT), desired graph density (D), directed flag (DF) which identifies directed graphs, and the time frame (T) in case of a dynamic graph, which identifies number of timestamps the generated edges are assigned to, and structure preference (S). Note that this generator will generate repeated edges and it is up to the user to use multiple edges to build a weighted graph or ignore the repeated ones. The partite flag indicates whether or not nodes of the same type are allowed to connect to each other. This parameter allows our generator to create graphs commonly known as bipartite but have more than 2 node types. A node type is implemented by preparing a prefix for each type and appending it to the beginning of each node. T was introduced by RTG to create dynamic graphs where each group of edges are assigned to a time stamp.

Structure preference was also introduced in RTG to match the real-life graph property of having community structures, meaning that nodes form groups and possibly groups of groups. These groups are usually created by connecting similar nodes together to form a community. The structure preference, $S \in [0,1]$, boosts the probability of generating similar nodes to be connected such that the probability of generating a character a in the label of both nodes is $P(a,a) = p - p^2 \times (k-1) \times S - p \times q \times S$ while the probability of generating 2 different characters a and b is $P(a,b) = p^2 \times S$. Note that k is number of characters in the model, p is probability of choosing a character to be in the label, and q is the probability of generating the separator character.

Our generator is efficient (see Fig. 2) and was successfully used to generate graphs with billions of edges.

5   WGB in Action

It has been shown that the probability of all generated words (nodes) with i characters is higher than the probability of all words with $i+1$ characters [20]. In other words, if the degree of each node is computed and all generated nodes are sorted alphabetically based on their labels, then nodes with fewer characters are expected to have a higher degree. There are only k possible words with one character, k^2 possible words with two characters, and k^n possible words with n characters.

This property could be used when generating real-life graphs. For example, given a list of website names, we could generate a sample of the web graph by categorizing these websites into multiple categories based on their expected popularity. Then popular websites are assigned to nodes with one character and less popular websites are assigned to nodes with two characters and so on. The same approach could be used to generate a social network or any other graph given a sample of the most important entities (nodes) in that graph.

The same property could be used to customize the selectivity of online and update queries discussed in Sect. 3 by querying nodes with short labels to get high selectivity or nodes with long labels to aim for low selectivity. This is not relevant for iterative queries because they access the complete graph at least once. The workload queries do not have specific parameters. There is a non-trivial relation between generated data and the selectivity of each query. Given a dataset, the selectivity of a query depends on its parameter in the "where" clause. For example, a 1-hop neighbour query from a node with one character (high degree) has a high selectivity relative to the same query from a node with many characters. Using this approach, the parameters of the proposed workload queries could be customized to offer a diverse range of selectivity.

6   Graph Processing and Analysis Systems

Distributed graph systems in literature could be divided into two main categories: MapReduce-based systems and vertex-centric processing systems. We will analyze systems in both categories using WGB. The systems that we plan to analyse in the first category include Hadoop (http://hadoop.apache.org), Pegasus [22] and Haloop [23]. Those in the second category include Distributed GraphLab [4] and Pregel [3].

6.1   MapReduce-Based Systems

MapReduce [24] is a distributed data processing framework whose fundamental idea is to simplify parallel processing using a distributed computing platform that offers two interfaces: map and reduce. Programmers can implement their own map and reduce functions, while the system is responsible for scheduling and synchronizing the map and reduce tasks. MapReduce model can be used to solve the embarrassingly parallel problems[7], where little or no effort is required to partition a task into a number of parallel but smaller tasks. MapReduce is being used increasingly in applications such as data mining, data analytics and scientific computation. It has indeed been used by Google to implement Pagerank.

---

[7] http://parlab.eecs.berkeley.edu/wiki/media/patterns/map-reduce-pattern.doc

Despite the scalability and simplicity of MapReduce systems, the model has a few shortcomings in processing graph analysis algorithms. First, the data are distributed randomly in multiple partitions at every iteration. Graph analysis algorithms usually execute several iterations before the computation converges. Second, map and reduce functions should be executed in all partitions before the next iteration starts. This introduces a significant delay because there may be one partition that takes too long to finish, and this would delay the processing of all other partitions. Finally, it does not consider the sparsity of a graph. If only one data partition did not converge, the framework will consider all data partitions for another computing iteration instead of focusing on the non-converging data partition only.

Despite these shortcomings, we plan to evaluate the most popular open source MapReduce implementation, Hadoop, as a baseline for system evaluations. Part of the reason is that, all other MapReduce-based systems have compared themselves only to Hadoop. Although, their reported results demonstrate their superiority over Hadoop, each one uses a different version of Hadoop and sometimes even different hardware. Moreover, each system uses a different dataset with different sizes and different properties. Finally, each of these use different workloads. Our objective is to systematically compare these systems against a single baseline.

HaLoop and Pegasus were proposed to overcome some of Hadoop's problems. Pegasus is a matrix multiplication application implemented using Hadoop. The main proposed enhancement is changing how the data are stored and clustered. This enhancement decreases the data size and reduces the number of iterations to reach convergence.

HaLoop is a MapReduce variant developed specifically for iterative computation and aggregation. In addition to the map and reduce functions, HaLoop introduces AddMap and AddReduce functions to express iterative computation. To test the termination condition, the SetFixedPointThreshold, ResultDistance, SetMaxNumOfIterations functions are defined. To distinguish the loop-variant and loop-invariant data, AddStepInput and AddInvariantTable functions are introduced.

To avoid unnecessary scans of invariant data, HaLoop maintains a reducer input cache storing the intermediate results of the invariant table. As a result, at each iteration, the mappers only need to scan the variant data generated from the previous jobs. The reducer pulls the variant data from the mappers, and reads the invariant data locally. To reduce the time to reach a fixed point, the reducer output cache is used to store the output locally in the reducers. At the end of each iteration, the output that has been cached from the previous iteration is compared with the newly generated results to detect whether the fixed point is reached. To assign the reducers to the same machine in different iterations, the MapReduce scheduler is also modified.

6.2   Vertex-Centric Systems

Systems in the vertex-centric category are based on a simple idea: "Think as a vertex". These systems require the programmer to write one compute function that a vertex executes at every iteration. The compute function operates on each vertex and may update the vertex state based on its previous state and the message passed to it from the preceding vertices. As noted earlier, the iterations stop when the computation converges to a fix point.

The first system proposed in this category was Pregel [3] and is used by Google to execute the Pagerank algorithm. In Pregel, data are partitioned across multiple computing nodes, and each node picks a graph vertex in its partition to start executing the compute function. A node may decide not to participate in future iterations (which reduces the load) and can communicate with other vertices through messages. The system is synchronized in the sense that all computing nodes need to complete executing their compute function before the subsequent iteration can start.

The architecture of Distributed GraphLab [4] is similar to Pregel while the processing model is different: a user-defined update function modifies the state of a vertex based on its previous state, the current state of all of its adjacent vertices, and the value of its adjacent edges. An important difference between GraphLab and Pregel is in terms of their synchronization model, which defines how the vertices collaborate in the processing. Pregel only supports Bulk Synchronization Model (BSM) [25] where, after the vertices complete their processing of a iteration, all vertices should reach a global synchronization status, while GraphLab allows three choices: (fully or partially) synchronized and asynchronized.

7   Conclusion and Future Work

In this paper we proposed a new benchmark, WGB. WGB has a data generator built on top of the RTG generator, which is efficient and user friendly. In this generator we solved one of the main issues in the RTG generator by hiding the input parameter that is not related to graph properties. We also introduced more parameters to enable more flexibility. WGB workload specification includes three categories of graph queries: online, update, and iterative queries. As discussed, most graph applications will use combinations of the graph queries we introduced.

In the future, we are planning to use WGB to perform the first systematic experimental study of parallel/distributed graph systems such as Hadoop Map-Reduce [24], Pegasus [22] and HaLoop [23], Distributed GraphLab [4], and Pregel [3]. It is also interesting to investigate how WGB might be integrated with BigBench [5] by generating a graph that represent the relationships between customers, products and retail stores such that the properties of these relationships match real-life graph properties.

Acknowledgments. This research was partially supported by a fellowship from IBM Centre for Advanced Studies (CAS), Toronto.

References

1. Dominguez-Sal, D., Martinez-Bazan, N., Muntes-Mulero, N., Baleta, P., Larriba-Pay, J.L.: A discussion on the design of graph database benchmarks. In: Proceedings of 2nd TPC Technology Conference on Performance Evaluation, Measurement and Characterization of Complex Systems, pp. 25–40 (2011)
2. Ciglan, M., Averbuch, A., Hluchy, L.: Benchmarking traversal operations over graph databases. In: Proceedings Workshops of 28th International Conference on Data Engineering, pp. 186–189 (2012)
3. Malewicz, G., Austern, M.H., Bik, A.J., Dehnert, J.C., Horn, I., Leiser, N., Czajkowski, G.: Pregel: a system for large-scale graph processing. In: Proceedings of 2010 ACM SIGMOD International Conference on Management of Data, SIGMOD '10, pp. 135–146 (2010)
4. Low, Y., Gonzalez, J., Kyrola, A., Bickson, D., Guestrin, C., Hellerstein, J.M.: Distributed graphlab: a framework for machine learning in the cloud. Proc. VLDB Endow. **5**(8), 716–727 (2012)
5. Ghazal, A., Rabl, T., Hu, M., Raab, F., Poess, M., Crolotte, A., Jacobsen, H.-A.: Bigbench: towards an industry standard benchmark for big data analytics. In: Proceedings of ACM SIGMOD International Conference on Management of Data, pp. 1197–1208. ACM (2013)
6. Wang, L., Zhan, J., Luo, C., Zhu, Y., Yang, Q., He, Y., Gao, W., Jia, Y., Shi, Y., Zhang, S., et al.: Bigdatabench: A big data benchmark suite from internet services (2014). arXiv preprint arXiv:1401.1406
7. Leskovec, J., Chakrabarti, D., Kleinberg, J., Faloutsos, C., Ghahramani, Z.: Kronecker graphs: an approach to modeling networks. J. Mach. Learn. Res. **11**, 985–1042 (2010)
8. Ming, Z., Luo, C., Gao, W., Han, R., Yang, Q., Wang, L., Zhan, J.: BDGS: a scalable big data generator suite in big data benchmarking (2014). arXiv preprint arXiv:1401.5465
9. Appel, A.P., Faloutsos, C., Junior, C.T.: Graph mining techniques: focusing on discriminating between real and synthetic graphs. Bioinformatics: Concepts, Methodologies, Tools, and Applications, vol. 3, pp. 446–464. Information Resources Management Association, USA (2013)
10. Bizer, C., Schultz, A.: The Berlin SPARQL benchmark. J. Seman. Web Inf. Syst. **5**(2), 1–24 (2009)
11. Morsey, M., Lehmann, J., Auer, S., Ngonga Ngomo, A.-C.: DBpedia SPARQL benchmark – performance assessment with real queries on real data. In: Aroyo, L., Welty, C., Alani, H., Taylor, J., Bernstein, A., Kagal, L., Noy, N., Blomqvist, E. (eds.) ISWC 2011, Part I. LNCS, vol. 7031, pp. 454–469. Springer, Heidelberg (2011)
12. Guo, Y., Pan, Z., Heflin, J.: LUBM: A benchmark for OWL knowledge base systems. Web Seman.: Sci. Serv. Agents World Wide Web **3**(2), 158–182 (2005)
13. Schmidt, M., Hornung, T., Lausen, G., Pinkel, C.: SP²1 SPARQL performance benchmark. In: Proceedings of 25th International Conferrence on Data Engineering, pp. 222–233 (2009)

14. Duan, S., Kementsietsidis, A., Srinivas, K., Udrea, O.: Apples and oranges: a comparison of RDF benchmarks and real RDF datasets. In: Proceedings of ACM SIGMOD International Conference on Management of Data, pp. 145–156 (2011)
15. Aluç, G., Özsu, M.T., Daudjee, K., Hartig, O.: Chameleon-db: a workload-aware robust RDF data management system, University of Waterloo, Technical report, CS-2013-10(2013)
16. Yu, J., Cheng, J.: Graph reachability queries: a survey. In: Aggarwal, C.C., Wang, H. (eds.) Managing and Mining Graph Data. Advances in Database Systems, vol. 40, pp. 181–215. Springer, Heidelberg (2010)
17. Spillane, S.R., Birnbaum, J., Bokser, D., Kemp, D., Labouseur, A., Olsen, P.W., Vijayan, J., Hwang, J.-H., Yoon, J.-W.: A demonstration of the G* graph database system. In: 2013 IEEE 29th International Conference on Data Engineering (ICDE), Los Alamitos, CA, USA, pp. 1356–1359. IEEE Computer Society (2013)
18. Aggarwal, C.C., Wang, H.: A survey of clustering algorithms for graph data. In: Aggarwal, C.C., Wang, H. (eds.) Managing and Mining Graph Data. Advances in Database Systems, vol. 40. Springer, Heidelberg (2010)
19. Akoglu, L., Faloutsos, C.: RTG: a recursive realistic graph generator using random typing. Data Min. Knowl. Disc. **19**(2), 194–209 (2009)
20. Miller, G.A.: Some effects of intermittent silence. Am. J. Psychol. **70**(2), 311–314 (1957)
21. Leskovec, J., Kleinberg, J., Faloutsos, C.: Graph evolution: densification and shrinking diameters. ACM Trans. Knowl. Discov. Data, **1**(1), Article 2, pp. 1–41 (2007)
22. Kang, U., Tsourakakis, C.E., Faloutsos, C.: PEGASUS: a peta-scale graph mining system implementation and observations. In: Proceedings of IEEE International Conference on Data Mining, 2009, pp. 229–238 (2009)
23. Bu, Y., Howe, B., Balazinska, M., Ernst, M.D.: The HaLoop approach to large-scale iterative data analysis. VLDB J. **21**(2), 169–190 (2012)
24. Dean, J., Ghemawat, S.: MapReduce: simplified data processing on large clusters. In: Proceedings of 6th USENIX Symposium on Operating System Design and Implementation, pp. 137–149 (2004)
25. Valiant, L.G.: A bridging model for parallel computation. Commun. ACM **33**(8), 103–111 (1990)

HcBench: Methodology, Development, and Full-System Characterization of a Customer Usage Representative Big Data/Hadoop Benchmark

Vikram A. Saletore[1]([⊠]), Karthik Krishnan[2], Vish Viswanathan[1], and Matthew E. Tolentino[1]

[1] Data Center Group and Software Services Group, Intel Corporation, DuPont, USA
{vikram.a.saletore, vish.viswanathan,
matthew.e.tolentino}@intel.com
[2] Amazon Web Services, Amazon.com Inc., Seattle, USA
kkr@amazon.com

Abstract. The Hadoop platform for Map-Reduce is extensively for Big Data batch analytics as well as interactive applications in e-commerce, telecom, media, retail, social networking, and other areas. However, to date no industry standard benchmarks exist to evaluate the true performance of a Hadoop cluster.

Current Hadoop benchmarks such as HiBench, Terasort, etc. in the open source domain fail to capture the real usages and performance of a Hadoop cluster in a datacenter. Given that typical Hadoop deployments process jobs under strict Service Level Agreement requirements, benchmarks are needed to evaluate the effects of concurrently running such diverse analytics jobs for performance comparison and cluster configuration.

In this paper, we present the methodology and the development of a customer usage representative Hadoop benchmark which includes a mix of job types, variety of data sizes, with inter-job arrival times as in a typical datacenter. We present the details of this benchmark and discuss application level, micro-architectural and cluster level performance characterization on an Intel Sandy Bridge Xeon Processor Hadoop cluster.

Keywords: Big Data · Hadoop benchmark · Cluster performance · Workload Characterization · Map-Reduce · Performance modeling

1 Introduction

With the advent of the Map-Reduce paradigm from Google [1] initially developed for Internet search; its open source version Hadoop data management framework now enjoys a credible and large ecosystem for Big Data analytics. Hadoop has fueled and simplified Map-Reduce style of programming over large clusters of commodity hardware for scalable and cost-effective solution for analytics applications. Map-Reduce Hadoop has also been adopted by a large number of organizations for

© Springer International Publishing Switzerland 2014
T. Rabl et al. (Eds.): WBDB 2013, LNCS 8585, pp. 73–93, 2014.
DOI: 10.1007/978-3-319-10596-3_7

ETL (Extract, Transform, and Load), web log and click stream analytics, graph construction, and other major IT applications for business value.

Although the Map-Reduce paradigm was initially used for batch processing of large-scale data analytics, it is now routinely being used for processing a large number of short interactive query jobs similar to traditional RDBMS [2]. Employing a large cluster for both interactive queries and batch jobs improves the utilization and efficiency of cluster hardware's compute, storage I/O, and network resources as long as the datacenter operations team can meet their customers' response times and throughput SLAs. Unlike traditional High Performance Computing (HPC) clusters, Map-Reduce/Hadoop clusters are increasingly used as time-shared systems where many diverse types of jobs may be running concurrently on any set of cluster nodes [2, 3].

A key challenge many organizations now face is how to size, configure, deploy, optimize, measure, and evaluate the true performance of a Map-Reduce/Hadoop cluster given the large diversity of potential jobs and workloads that may be scheduled to run concurrently by different users. Even if a single job is known to be I/O intensive in isolation, when that job is scheduled to run at the same time as many other different jobs, does the storage or network I/O remain the bottleneck? Are the performance dependencies of current micro-benchmarks reflected at scale in a multi-tenancy environment? Although there are some recent industry wide attempts [4] to develop a standard Hadoop benchmark, most use the traditional approach of running single application in isolation. While useful, the characteristics of a single application in isolation are not reflective of today's multi-tenant Hadoop deployments.

What is currently available in the open-source community are micro-benchmarks, such as HiBench [5], PigMix [6], GridMix3 [7], etc. that enables one to stress-test different aspects of the cluster. However, these micro-benchmarks represent narrow functional slices of customer's applications. While stress testing of Hadoop cluster is important for identifying performance bottlenecks, the micro-benchmarks fail to represent or capture customer's datacenter usages or measure the true performance of a production cluster.

Rather than leveraging simple micro-benchmarks in isolation, there have been efforts to capture some cluster-level characteristics of commercial deployments. For example, SWIM [9] uses a collection of abstracted customer workload traces from large data centers to generate synthetic benchmarks that can be replayed on a smaller test cluster. However, it has several drawbacks. First, it does not capture or represent real analytics computations and thereby does not provide insight into the compute-bound processing aspects of Hadoop workloads [8]. Instead it focuses on the impact of I/O movement of data within the cluster. Second, SWIM also has issues with its data size scaling methodology to benchmark smaller clusters. This is because when a large percentage (>90 %) of customer jobs operates on small (10s of MB) data sizes [3, 9] in large data centers, one cannot simply scale down the data size for a small test cluster. The resulting SWIM benchmark would then require very little compute and very low storage and network traffic. These shortcomings in fact, make SWIM benchmarks less-representative of actual customer workloads.

Our contributions in this paper are as follows:

- Our methodology and the benchmark captures the following attributes of customer representative usages; diverse mix of real analytic compute, storage I/O, and networking, multi-tenancy and concurrent processing of Hadoop jobs, varied input date sizes, inter-job arrival pattern, and performance repeatability.
- Hadoop clusters are typically architected and configured primarily with storage bottlenecks in mind. Based on the characterization of our proposed benchmark we observe that customer representative usage workloads also tend to be compute bound. Unlike SWIM benchmarks, our benchmark includes real analytic compute requirements.
- The benchmark also enables data center operators to configure, optimize, and evaluate Hadoop cluster and measure its true performance. In addition, this benchmark can also be used measure and compare different Hadoop systems under realistic customer usages.
- The micro-architectural level characterization of this benchmark enables one to develop performance estimation models for the current and potentially future generation CPU and system architecture.

The paper is organized as follows. In Sect. 2 we discuss the overview of prior related work. Section 3 presents the methodology for the development of our benchmark. In Sect. 4 we discuss the application level performance of HcBench. In Sect. 5 and Sect. 6 we discuss OS and micro-architectural level performance characterization respectively. We present our conclusions and future work in Sect. 7.

2 Related Work

In this section we discuss the state of the current benchmarks to evaluate Hadoop cluster performance. We also discuss why benchmark characterization is useful for identifying system bottlenecks. HiBench 2.0 [5] in the open source domain consists of a suite of 11 Hadoop micro-benchmarks that represents a wide diversity of individual applications but fails to capture job mixes, varied data sizes, and job arrival rates representative of customer usages. PigMix [6] consists of 12 queries to test performance of Pig for large data sizes. Although it has a good collection of queries, these are not representative of Pig deployments, data sizes, query mix, and query arrival rates usages in customer deployments. Hive Benchmark [10] uses the data warehousing built over Hadoop and uses five queries derived from [11] similar to one used in parallel databases. However, Hive benchmark fails to capture varied data sizes, and arrival rates representative of production usage of Hive. GridMix3 [7] takes as input a job trace derived from job logs collected via Rumen [12] on a cluster and constructs a synthetic job with similar byte and record patterns. The synthetic job is replayed to generate comparable load on the I/O subsystem as intended in the original production trace. However, it does not model CPU and memory usages and job dependencies in the job mix.

Workload Characterization at Taobao [13] brings out interesting aspects of customer usages on a large production Hadoop cluster. Yunti is a 2000 node Hadoop cluster at Taobao where they have characterized workload traces collected for about

1 Million jobs over a period of about 2 weeks. Their findings show that jobs arrive at a fairly regular pattern, and pre-emptions are caused by high priority jobs frequently. Small jobs constitute a majority of jobs while medium and large jobs take up a fair share. 50 % of jobs access and process less than 64 MB data from HDFS (Hadoop Distributed File System). 80 % of jobs also write less than 64 MB. Shuffle data causes high network traffic while CPU utilization fluctuates and correlates to job arrival rates.

SWIM [2, 9] synthetic benchmark was developed by UC Berkeley based on collection of production Hadoop traces in collaboration with Facebook and Cloudera. The SWIM methodology and benchmark uses sampling technique to abstract trace data. Also, as discussed earlier, SWIM workloads tend to be dominated by IO bound jobs and do not include any real analytic processing. SWIM fails at representing compute jobs [8]. Since it is not possible to accurately compare compute bound phases using SWIM, one could conceivably purchase and configure a Hadoop cluster with less powerful CPU server platforms resulting in failing to meet customer SLAs. Also, the SWIM methodology of scaling down data size for smaller clusters for benchmarking has issues. One cannot simply scale down job data sizes running on very large clusters (say 1000s of nodes) [9] to benchmark say a small 10 node cluster. E.g. if we were to scale down small jobs data size (average data size 871 KB in [3] for 90 % of customer usages) as in Facebook's 600 node cluster [3], it will result in a scaled down job data size of ~ 9 KB in the SWIM benchmark for a 6-node test cluster. Such a benchmark would require very little compute and very small storage traffic severely underestimating Hadoop cluster requirements and performance.

We believe that to measure and characterize the real performance of a Hadoop cluster, a benchmark must comprehend realistic compute, storage and networking aspects of a Hadoop cluster.

3   Methodology for Benchmark Development

In this section we present the development of this benchmark and its features. One of the primary goals of a benchmark is, that it should be able to stress system resources to its maximum performance capabilities for a given usage scenario. For example, TPC-H is a decision support benchmark from TPC.org [14, 15] and is used to measure performance with typical ad hoc data warehouse queries. Benchmark studies are also expected to reveal system bottlenecks (Software, OS, CPU, storage, network, memory, others, etc.) for a given customer usage, which the data center teams can later address to configure and optimize hardware and software to achieve better performance. We discuss the attributes of this benchmark in detail in the next few sections.

Based on studies of customer datacenter Hadoop cluster usages in the analytics space [2, 13] we have developed the Hadoop benchmark with features such as: Jobs diversity, Large number of jobs, Variety of input data sizes, Real computation, storage I/O, and network intensive, Inter-job arrival times, and Response Time SLA vs Job Throughput, easy to run, and completes in a reasonable amount of time. We will go over each of these features in the following sections.

3.1 Job Diversity

Let's consider job diversity in a typical Hadoop cluster. Unlike an HPC cluster, the Hadoop cluster hosts a variety of applications running concurrently in a multi-tenant or multi-programming mode. In datacenters, it is very common to utilize the Hadoop cluster as a time-sharing system for large number of jobs and job types [2] that include a mix of e-commerce, telecom, media, and retail workloads. The Yunti cluster at Taobao [13] processes a very diverse set of jobs from numerous applications such as commodities recommendations, traffic statistics, advertisement analytics, and user behavior analysis.

The current job mix in HcBench includes Telco-CDR (Call Data Records) interactive query workload, Hive Log Query jobs, K-Means Clustering iterative workload in machine learning, and Terasort jobs as shown in Table 1. Except for the Telco-CDR job, all other workloads are available in the open source domain [16]. Although we have included only a small mix of Hadoop jobs, our benchmark framework is flexible and these jobs can be augmented with other customer specific usages.

The Telco-CDR queries contain 5 different queries (e.g. cell tower usages, WAP fees, cell phone plan limits, etc.) applied to synthetic data. Each query results in multiple Map-Reduce jobs.

Hive Query selects PageRank URLs, aggregates by source, average of PageRank, etc. The application results in the creation of tables for ranking and user visits and performs inserts and joins, also resulting in several Map-Reduce jobs.

K-Means algorithm used in many Machine Learning applications clusters data that are similar to each other on synthetic data using the Mahout library. K-Means is executed with 4 iterations for convergence.

We used Terasort as a proxy for ETL and transform jobs to sort a large data set of random 100 byte records.

3.2 Number of Jobs

We set out with the goal of a large number of job submissions that completes in a reasonably short amount of time. Since each job submission results in an average of 4–5 Map-Reduce jobs, the total number of Hadoop jobs executed is about 4.5X that of

Table 1. HcBench benchmark data set mix

Job Types	Prob of Job Selection	Workload	Input Data Sizes	Data Size	Prob of Data size Selection
1	33%	Telco Queries (CDR)	128MB to 4096MB	Small	80%
		Interactive	8GB, 16GB, 32GB	Large	20%
2	33%	Log Hive Queries	128MB to 4096MB	Small	80%
		Interactive	8GB, 16GB, 32GB	Large	20%
3	33%	K-Means Clustering	128MB to 4096MB	Small	80%
		Iterative	8GB, 16GB, 32GB	Large	20%
4	1%	Terasort (Transform)	64GB	Very Large	100%

submissions. For HcBench benchmark we use 101 job submissions; 100 jobs submission come from Telco-CDR, Hive, and K-Means job submissions with 1 Terasort job representing large transform job.

3.3 Input Data Size Mix

It has been shown earlier in [2, 13] that majority (80–90 %) of interactive jobs operate on small data sets with a small percentage of aggregate and transform jobs on medium and large data sets. Since the size of the input data varies with customer usages, we used 9 unique data sizes for each job in HcBench as shown in Table 1; data of sizes 120 MB–4096 MB are considered as small and 8 GB–32 GB as large. Multiple data sizes within each set of "small" and "large" are generated in powers of 2.

3.4 Compute, Storage, and Network Intensive Jobs

Unlike other synthetic benchmarks which do not include any analytic processing jobs, HcBench benchmark includes real analytic computations. The interactive and log queries are compute bound during Map stage. K-Means jobs are compute bound in iteration and I/O bound in clustering. Terasort is compute bound in Map stage, storage in shuffle stage, and network bound in Reduce stage.

3.5 Inter-job Arrival Pattern

Analysis of sampled trace data collected on large Hadoop clusters at customer data centers [9] shows that the inter-job arrival times can be approximated by Gamma Probability Density Function (PDF) [17]. For HcBench we use Gamma PDF and populate with random 101 submissions to obtain resulting inter-job arrival times as shown in Fig. 1. Although, HcBench has only 101 jobs, the general trend and distribution matches fairly well to the large data sets in [9].

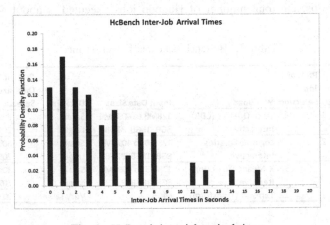

Fig. 1. HcBench inter-job arrival times

3.6   Putting It All Together

We now describe how we put all the above features together to develop the HcBench benchmark. We have developed our methodology in Java, Python and shell that takes into account of all the features to generate a static schedule of all the 101 job submissions for HcBench. The static schedule is a list of commands that specifies the inter-job arrival times followed with a job invocation command to execute a specific analytic job to operate on a specific data size.

Consider job diversity. Given the 4 job types (Telco-CDR, Hive Queries, K-Means, and Terasort), we assign a probability of selection to each job type as shown in Table 2. For our benchmark, we assume that the probability of selecting an interactive query, an ad hoc query, or an iterative job is equally likely. Now for each job type, we generate 9 unique data sizes in HDFS. Those include a set of small data sizes in powers of 2 between 128 MB and 4096 MB plus another set of large data set of sizes between 8 GB and 32 GB. In addition, the data sets are assigned weights such that the probability of selecting any small dataset size is 80 % and the probability of selecting a large dataset size is 20 %. Since the benchmark also includes a large transform (Terasort as a proxy) job, its 64 GB size dataset size is given a probability of 100 % and selected only once.

Table 2. Job submission schedule

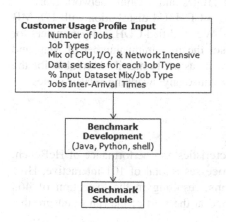

| /home/HcBench/telco-cdr/run.telco.sh telco-cdr_4GB_table 2 & |
| sleep 2 |
| /home/HcBench/hive-query/run-hive.sh   hdfs://nnode/input/Hive/input-comp.128MB & |
| sleep 0 |
| /home/HcBench/telco-cdr/run.telco.sh telco-cdr_512MB_table 1 & |
| sleep 8 |
| /home/HcBench/hive-query/run-hive.sh   hdfs://nnode/input/Hive/input-comp.2GB & |
| .... |
| Sleep 2 |
| /home/HcBench/terasort/run.tsort.sh /input/terasort/input.64GB & |
| .... |

Fig. 2. Benchmark methodology

Figure 2 shows our methodology for developing this schedule. Our benchmark methodology takes as its input the attributes of a customer usage profile (number of jobs, job types, data sizes, probability of selection, inter-job arrival times, job submission command, etc.). This data is processed and the resultant inter-job time and the command to submit a specific job is recorded to a multi-job schedule shell script. The schedule for HcBench can then be invoked to execute on a Hadoop cluster. It is assumed that the necessary datasets needed for all job submissions is stored in HDFS (Hadoop Distributed File System) prior to running the benchmark schedule script.

Table 2 shows a sample schedule for HcBench. Execution of this static schedule is essentially the execution of this benchmark. The schedule also ensures the repeatability of benchmark with minor variations in total execution times as we will show later.

4 HcBench Application Level Performance

4.1 Hadoop Cluster HW and SW Components

We have collected detailed performance characterization data on our proposed benchmark on an 8-Node Hadoop test cluster (Fig. 3). Each data node in the cluster is configured with 2 Intel® Xeon® Processors (32 threads with Hyper-Threading enabled) Sandy Bridge platforms, 64 GB Memory, 1 480 GB SSD OS drive and 3 480 GB SSDs 6 Gbit/s SATA drives for HDFS data, and 10 Gbit Intel® NIC. The ~11.5 TB of storage capacity is sufficient to store up to 3.5 TB HDFS input data with 3X replication. We used SSDs for HDFS storage to have sufficient storage bandwidth with fewer numbers of drives and providing sufficient capacity for our benchmark data. Also, we used 10 Gbit Ethernet fabric so that we are not network bandwidth limited for a cluster of this size. We understand that this may not represent a real-world Hadoop cluster. However, we used flash drives and the 10 Gbit fabric for experimental purposes as we can modify the configurations and use rotating HDDs and 1 Gbit/s network fabric.

For the Hadoop framework we used Cloudera CDH3U4 distribution with 128 MB HDFS block size and enabled its fair-scheduler. We modified CDH3U4 parameters for optimized performance. In addition, we also set the maximum number of Map and Reduce tasks per node to 24 and 16 respectively, so as not to heavily oversubscribe the cluster memory (64 GB per node) resources to cause any swapping related issues.

4.2 Benchmark Performance Summary

Table 3 lists the high level performance characteristics and performance of HcBench. The benchmark runs for about 24 min and processes a total of 101 interactive, Hive query, iterative, and transform job submissions, resulting in the execution of 465 Hadoop jobs. The buffer page caches are flushed at the start of the run to ensure that data is read fresh from HDFS.

Table 3. HcBench perf. characteristics

Number of Job Types	4
Job Types	Interactive, Hive, Iterative, Xform
Num. of Job Submissions	101
Total Num. of Map-Red Jobs	465
Max Number of Data Sizes/Job	9
Small Data Sizes (MB)	128, 256, 512, 1024, 2048, 4096
Large Data Sizes (GB)	8, 16, 32
Large Transform Data Size	64GB
Inter-Job Arrival Distribution	Gamma Prob. Density Function
Average Job Response Time	396 seconds (with Queue Delays)
Average Map Time	30.04 seconds
Average Reduce Time	13.34 seconds
Total HDFS Bytes Read	792GB
Total Running Time	1433 seconds (23.8 minutes)
Avg. Data Throughput MB/Sec	552 MB/sec
Avg. Jobs Throughput Jobs/Hr	253.7 Jobs/Hr.

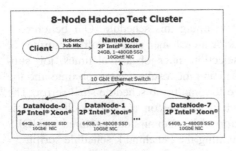

Fig. 3. 8-Node Hadoop test cluster

4.3 Benchmark Performance Summary

Table 3 lists the high level performance characteristics and performance of HcBench. The benchmark runs for about 24 min and processes a total of 101 interactive, Hive query, iterative, and transform job submissions, resulting in the execution of 465 Hadoop jobs. The buffer page caches are flushed at the start of the run to ensure that data is read fresh from HDFS.

4.4 Data Throughput

A key performance metric that is of interest for Hadoop jobs processing is the rate at which HDFS data is read and processed. Figure 4 shows HDFS data throughput for HcBench. This data is easily obtained with post processing of Hadoop job logs. The throughput shown is a running average measured every second for a window of past 1 min. We observe that HcBench is capable of processing at an average of 552 MB/s and with a peak rate of about 3 GB/s of data from the HDFS.

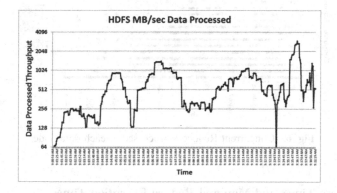

Fig. 4. HDFS input data processed throughput

4.5 Concurrency During Map and Reduce Phases

Concurrency of map tasks and reduces tasks, captured per second across all data nodes are shown in Figs. 5 and 6. Data shows that during Map and Reduce phases, there is a high degree of concurrency except during ramp-up and ramp-down phases of the benchmark. Hadoop framework typically assigns 1 CPU thread per Map or Reduce task plus an additional thread to the task-tracker.

We observe that once past the ramp-up phase, there is a high degree of concurrency in processing of Map tasks and Reduce tasks from different Hadoop jobs. One would like to exploit maximum concurrency in processing of all the phases of Map-Reduce processing such that all the scheduled jobs make forward progress towards completion to meet respective customer SLAs. As indicated earlier, limiting the maximum number of Map and Reduce tasks allows forward progress during the Map, Shuffle, and Reduce phases and also avoids OS swapping issues and performance degradation for the given memory capacity per data node.

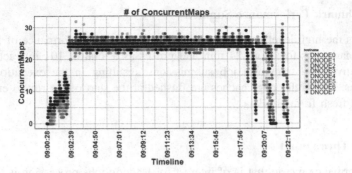

Fig. 5. Concurrent Maps processed at each data node

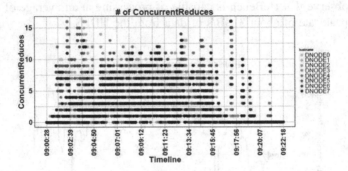

Fig. 6. Concurrent Reducers processed at each data node

4.6   Response Times and Map and Reduce Execution Times

Figure 7 shows measured response times for each of the 101 job submitted and processed. The response times include the waiting time spent in the job submission queue.

The measured average response time for jobs submitted is 396 s. We would like to reiterate again that on an average a single job submission results in the execution of about 4–5 successive Map-Reduce Hadoop jobs. As indicated earlier, 101 job submissions result in the execution of 465 jobs. Each time a Map-Reduce job is invoked; it is scheduled and processed along with Map-Reduce jobs from other scheduled submissions.

Figure 8 shows execution times (excluding queuing delays) for each of the 465 Hadoop jobs. Each bar in the figure shows execution times for Map task (in red), Map-Reduce tasks overlapped (in dash), and Reduce task (in black). Since Map and Reduce tasks from multiple jobs could be executing concurrently, no single job has exclusive access to the full cluster computing resources, resulting in potentially increased execution times for Map, Shuffle, and Reduce phases.

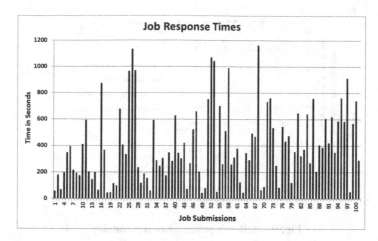

Fig. 7. Job Response Times (including queuing delays) for 101 submissions

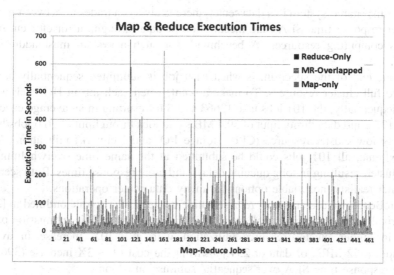

Fig. 8. Map-Reduce Task exe. Times (excluding queuing delays) for 465 Hadoop jobs (Color figure online)

4.7 Benchmark Repeatability

Another important criterion in a benchmark is its ability to show consistent and repeatable performance within a reasonable tolerance. Figure 9 shows different performance characteristics of HcBench over multiple runs. The data in the chart shows that the variation of performance metrics is less than 2.5 %. Given the wide variety of the job mix in the benchmark that includes compute, storage and network I/O in addition to OS interaction, it is reasonable to see a larger variation than other CPU-Memory compute intensive single application benchmarks.

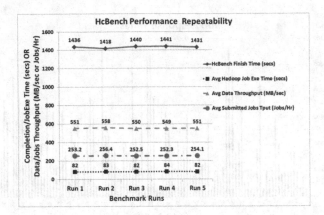

Fig. 9. HcBench performance repeatability

4.8   Response Time Job SLA vs Throughput

For any time shared system in a data center, there is always a trade-off between meeting customer response time SLAs and meeting job or data throughput for efficient use of cluster's computing resources. A benchmark for such a system must address this tradeoff.

At one end of the spectrum is when each job is submitted sequentially and has access to full cluster resources. To make a point, when each job in HcBench is submitted sequentially, the 101 jobs take 13663 s (3.8 h) resulting in an average execution time of 135 s and data throughput of 59.4 MB/s and jobs throughput of 26.6 jobs/h with extremely low cluster resource (CPU, storage I/O, and Network) utilization. At the other extreme, all 101 jobs could be submitted at the same time overwhelming the entire cluster, resulting in long queuing delays and large response times. Both scenarios are neither realistic nor viable options for many data center operations.

In HcBench we use customer representative usage similar to one outlined in [2] for inter-arrival times to simulate realistic job submission pattern. This submission pattern results in higher cluster resource utilization with a $\sim 10X$ increase in average throughput (552 MB/s of data or 254 jobs/h) at the cost of $\sim 3X$ increase (396 s) in average response time SLA over sequential submission of jobs.

5   HcBench OS Level Performance Characterization

In this section we discuss OS level performance characterization data collected using Linux sar and post processed via an internally developed system visualizer tool "sysviz" [18] written in language R. We discuss two types of charts in this section. One is the performance metric measured every second on each of the data nodes (in grey scale) during the execution of the benchmark. The other is the bar chart (also in grey scale) histogram data in buckets (powers of 2). The height of the bar chart in a specific bucket indicates the percentage of total runtime that the specific metric achieved on that data node. The histogram bar charts give us insight into the spread (maximum, median, and minimum) of a specific metric during the execution of the benchmark. We decided not to show average values they tend to hide the performance variation of a metric.

5.1 CPU Utilization

Figure 10 show user CPU utilization for all the data nodes collected during the benchmark execution. Once the jobs are submitted and start executing, the cluster CPU utilization quickly ramps up to 100 % with 90 %–95 % in user mode. Since multiple jobs are submitted, Map, Shuffle, and Reduce phases from different jobs are processed concurrently. We suspect that most of the CPU utilization in kernel mode is spent in disk and network I/O. Given that we observe almost 100 % CPU utilization, one simply cannot ignore the compute required for real analytics.

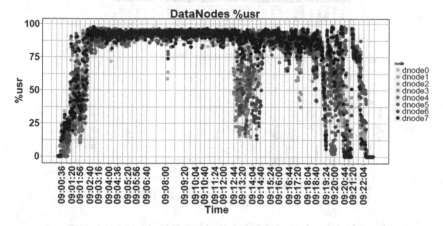

Fig. 10. User CPU utilization

We also observe a perturbation in utilization when a transform job starts up which requires significant storage accesses. As the last few jobs complete at the end, the CPU utilization drops as expected. This chart clearly shows that the benchmark is able to stress the compute nodes for maximizing performance. Hadoop CDH3U4 was configured with 1 GB/JVM. Data collected also indicate (not shown) that the memory consumption ramps up to about 80 % fairly quickly, stays high and peaks at 90 % before dropping off during the ramp-down phase at the end.

5.2 HDFS Disk Bandwidth and Request Size

Now, let us analyze storage accesses. Figure 11 (a) and (b) shows the HDFS read bandwidth measured every second during execution of HcBench. We observe that the read access bandwidth is not very uniform across all the data nodes. This is because with a large number (~ 80 %) of small jobs, the data sets can be covered by a few

128 MB HDFS blocks on fewer nodes resulting in non-uniform storage accesses across nodes. Also, this is fairly representative of real-world clusters with small data set sizes for many jobs. The histogram of read bandwidth in Fig. 11 (b) shows the spread of disk bandwidth across all the nodes. From the detailed data we observe disk bandwidths with a mean of 44.1 MB/s, median of 38.2 MB/s, std. dev. of 36.4 MB/s, and peak of 255.3 MB/s across the 3 SSD drives.

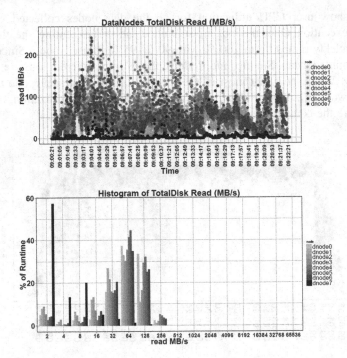

Fig. 11. (a) and (b): Total Disk READ bandwidth and histogram

Figure 12 (a) and (b) shows the write bandwidth and its histogram for the benchmark. We observe that the write bandwidth is also not uniform across data nodes and peaks at ~1 GB/s across the 3 SSD drives on a data node. We believe that this is due to the output of the large transform job written back to storage with 3X replication enabled in Hadoop. However, it is interesting to observe that about 50 % of the execution time, the Write bandwidth is actually fairly small around 3–4 MB/s as shown by the bar charts on the extreme left in the Fig. 12 (b). From the detailed data we observe bandwidths with a mean of 55.3 MB/s, median of 3.34 MB/s, std. dev. of 107.4 MB/s and a peak of 862.2 MB/s across the 3 SSD drives.

5.3   RX and TX Network Bandwidth

Let us now consider how this benchmark exercises the 10 GbE network fabric. Figure 13 shows Ethernet network receive bandwidth during execution. From the analysis of detailed data we observe a mean RX bandwidth of 0.41 Gbit/s and with a

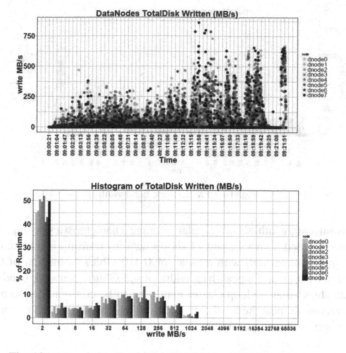

Fig. 12. (a) and (b): Total Disk WRITE bandwidth and histogram

std. dev. of 0.8 Gbit/s. The peak bandwidth reaches 7.7 Gbit/s for a small percentage of runtime. However, the RX bandwidth is not very uniform as it is mainly the intermediate shuffle data moving from many Map tasks to Reduce tasks. However, if we had configured the cluster with 1 Gbit/s Ethernet fabric, we would have limited the network bandwidth resulting in potentially lower performance.

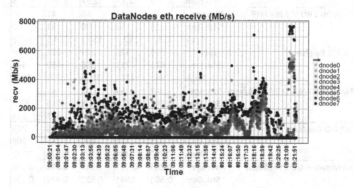

Fig. 13. RX network bandwidth

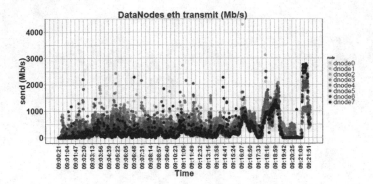

Fig. 14. TX network bandwidth

The transmit bandwidth shows similar average characteristics to that of receive bandwidth as discussed above. Figure 14 shows data for transmit bandwidth. The data shows the transmit bandwidth with a mean of 0.410 Gbit/s, a median of 0.3 Gbit/s, a std. dev. of 0.4 Gbit/s and a peak bandwidth of 4.3 Gbit/s. Network bandwidth data and its spread shows that Hadoop cluster may need to be configured with appropriate fabric infrastructure to meet the bandwidth required to achieve optimal performance for a customer usage workload profile.

Table 4. Performance summary across multiple runs

Performance		Run 1	Run 2	Run 3	Run 4	Run 5
User CPU Util (%)	Average	82.5	83.7	82.7	82.1	82.9
	Median	92.3	92.2	92.1	92.0	92.1
	Std. Dev.	20.8	19.2	21.7	21.2	21.3
System CPU Util (%)	Average	6	6	6	7	6
	Median	3	4	3	3	3
	Std. Dev.	10.0	11.8	9.6	11.8	10.2
Idle CPU Util (%)	Average	11	9	11	11	11
	Median	2.9	2.84	2.97	2.955	3
Total Disk RD I/O BW (MB/sec)	Average	40	42	40	41	40
	Std. Dev.	29	31	30	31	29
	Peak	113.6	112.3	120.8	114.5	103.3
RD I/O Req/sec	Average	295	325	289	291	294
	Std. Dev.	421	751	560	460	443
	Peak	2642	7132	3026	2235	2222
Total Disk WR I/O BW (MB/sec)	Average	49	53	48	47	51
	Std. Dev.	99.1	102.0	91.8	94.5	97.8
	Peak	560	563	434	571	573
WR I/O Req/sec	Average	276	298	267	278	276
	Std. Dev.	331	371	295	357	308
	Peak	1722	2385	1271	1374	1212
RX Network I/O BW (Gbps)	Average	0.289	0.307	0.318	0.292	0.309
	Std. Dev.	0.501	0.541	0.530	0.497	0.534
	Peak	4.881	6.675	5.167	4.673	4.406
TX Network I/O BW (Gbps)	Average	0.284	0.306	0.316	0.286	0.305
	Std. Dev.	0.280	0.302	0.300	0.271	0.298
	Peak	2.848	3.078	4.500	2.283	3.647

5.4   Performance Summary

Table 4 shows the summary of system performance characteristics of the benchmark over multiple runs. We show the mean, standard deviation, and peak values of the metrics to show the variation observed. The standard deviation and peaks give a good indication of hardware performance requirement for a Hadoop cluster configuration needed for these customer usages.

6   Micro-architectural Level Performance Characterization

In this section we focus on the micro-architectural aspects of the HcBench's performance characterization using EMon (Intel EMon software) tool. We first present the data on averages across the 8 data nodes of the Hadoop cluster. Later, we will focus on selected individual metrics.

6.1   Micro-architectural CPU Performance Metrics

Intel EMon tool allows us to read hardware performance monitoring events provided on the processor with time-based sampling. The EMon samples from performance counters is collected at every 200 ms (5 samples/s) and then post-processed for analysis using the EMon Data Processing (EDP), an internal Intel software analysis tool. Some of the key hardware events that were collected for further performance analysis are:

- CPU Utilization: Total system CPU utilization
- Kernel CPU Utilization: Kernel CPU utilization
- User CPI: Cycles Per Instructions in user code
- Kernel CPI: Cycles Per Instructions in kernel code
- LLC MPI: Last Level Cache Misses Per Instruction
- LLC Miss Latency: LLC Miss Latency in ns
- Reads Satisfied by Local or Remote DRAM
- Read and Write Memory Bandwidth
- Unhalted Clock Cycles: Number of cycles that the benchmark took for execution not counting when the CPU was halted
- CPU Instructions: Number of instructions retired by the CPU

6.2   Benchmark Full-Run: DataNode 0

In this section we present micro-architectural data collected for the full-run of the benchmark including the ramp-up in the beginning, steady-state in the middle, and the ramp-down at the end. We will focus here on the data collected datanode-0 of the Hadoop cluster. Data collected on other data nodes is very similar and we will discuss the overall average data for different CPU metrics.

Figure 15 shows the overall and Kernel CPU Utilization during the benchmark execution. This chart is very similar to the data shown in Fig. 10 earlier, although more accurate based on larger number of samples collected every 200 ms for about 6600 samples ($\sim$22 min of run time). CPU utilization in the kernel mode accounts for about 3–7 %.

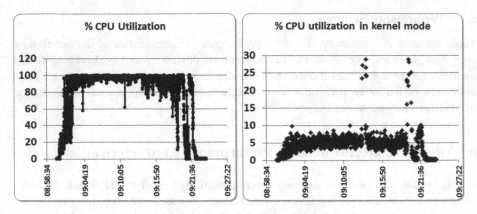

Fig. 15. CPU utilization for DataNode0

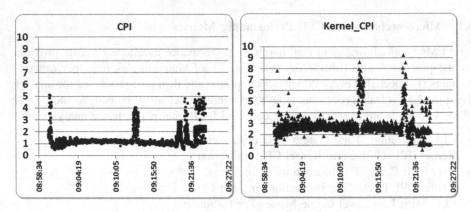

Fig. 16. Cycles Per Instruction (CPI) for DataNode0

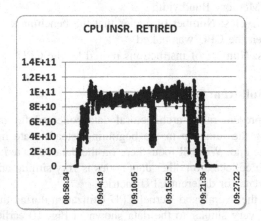

Fig. 17. LLC MPI and miss latency for DataNode0

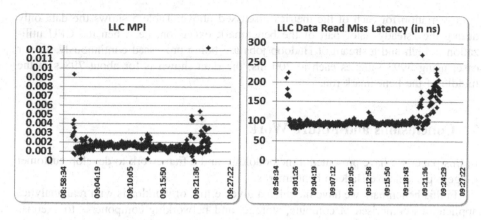

Fig. 18. LLC MPI and miss latency for DataNode0

Figure 16 shows the overall CPI of 1.1 and Kernel CPI of 2.8–3.0 for DataNode-0. The detailed measurement of CPI gives us a way to analyze the CPU performance using this benchmark. Figure 17 shows the CPU instructions retired per second. Figure 18 shows the last level cache misses per instruction and also data read miss latency in ns.

Detailed measurement of the CPU frequency, utilization, Instructions retired, CPI, LLC MPI, and LLC miss latency in processor cycles enables us to estimate the performance of the system. Based on detailed micro-architectural characterization including high-level OS and benchmark data, we can develop an analytical model for performance estimation. The analytical model also enables us to develop projection models for future generation platform and full system architecture.

6.3   Benchmark Run in Steady-State: DataNodes 0 to 7

Using metrics from a full benchmark run, including ramp-up and ramp-down phases may not give very accurate data to develop a performance estimation model as there is

Table 5. CPU metrics averaged across Hadoop cluster Data Nodes

Benchmark Steady-State(sample #1000-4992) for DataNodes 0-7	Average	Maximum	Minimum	Std. Dev
CPU frequency (in GHz)	3.29	3.30	3.26	0.016
CPU utilization (%)	96.6	97.0	95.6	0.440
CPU utilization in kernel mode (%)	7.7	9.5	5.0	1.482
User_CPI (Cycles Per Instruction)	1.12	1.16	1.09	0.026
Kernel_CPI	3.45	3.80	2.83	0.318
LLC MPI (misses per instruction)	0.0015866	0.0016637	0.0015093	0.000
Average LLC data read miss latency (in ns)	95	97	94	1.134
Reads satisfied by local DRAM (%)	63	65	61	1.368
Reads satisfied by remote DRAM (%)	34	37	33	1.453
Memory Read Bandwidth (MB/sec)	9,483	9,941	9,322	211.553
Memory Write Bandwidth (MB/sec)	7,057	7,445	6,858	182.006
Memory Total Bandwidth (MB/sec)	16,540	17,385	16,183	374.821
CPU_CLK_UNHALTED (clocks)	101,740,629,120	102,403,391,853	100,938,921,577	621,991,627
CPU Instructions Retired/sec	91,052,157,391	93,614,710,530	88,276,525,240	2,052,123,442

large variation for each of the metrics discussed above. Table 5 shows the data only during the "steady-state" part of the benchmark execution, i.e. when the CPU utilization is high and a stream of Hadoop jobs are being processed continuously. In our case, using 3993 samples each at 200 ms; the data shown is for about 798 s in the middle of the benchmark run.

7   Conclusions and Future Work

In this paper we have presented a methodology and a framework to develop customer usage representative Hadoop benchmarks.

In this benchmark we have included a diverse mix of workloads with real analytics applications comprised of compute, storage, and networking components to exercise different aspects of a representative data center Hadoop cluster. In addition, our benchmark also includes a number of scaled data sizes for each job in the mix and appropriately weighted to reflect customer data center usage. Furthermore, our benchmark also comprehends and simulates a data center representative inter-job arrival times. We have also demonstrated that the benchmark has other attributes that it completes in a reasonable time, shows repeatability and is easy to run.

This proposed HcBench enables us to measure and compare the true performance of a Hadoop cluster and identify potential system bottlenecks and evaluate cluster configurations. HcBench's performance metrics can also assist Map-Reduce performance studies for comparing the performance of Hadoop distributions. In addition, the benchmark also enables us to measure and analyze data to improve the performance of the Hadoop framework's software components.

Detailed performance characterization of this benchmark reveals interesting insights on the compute, storage, and network characteristics of a Hadoop cluster. For customer representative usages of interactive queries and transform jobs, results in high CPU utilization, fairly low storage I/O and not very high network utilization. Characterization data from this benchmark can be used to estimate hardware requirements and purchasing decisions for a Hadoop cluster. Furthermore, micro-architectural characterization data can also be used to develop analytical performance projection models for future generation cluster hardware components.

We are conducting further investigations to scale the benchmark as the size of the Hadoop cluster is scaled but at the same time represents the attributes of customer usages in the data center. Furthermore, we believe that we need to go beyond a single customer representative benchmark; as there are many diverse customer usages in the Hadoop ecosystem. Our framework enables customization of customer usage profiles for different application domains and usages in the Hadoop ecosystem. We plan to extend our work and develop a suite of customer representative benchmarks representing different customer usages. Lastly, once the improvements are in place, we plan to investigate to make this benchmark publicly available.

We are currently developing an analytical model using high-level OS level data and fine-grain level micro-architectural level CPU metrics. Once we validate this model, we plan to use simulation to develop performance projections using this benchmark for future generation server clusters.

Acknowledgements. Karthik Krishnan, now with Amazon Web Services at Amazon.com Inc. contributed to the significant development of this benchmark when he was at Intel. We would also like to thank our manager, Intel Fellow and Chief Server Architect of the Data Center Group, Dr. Faye Briggs for encouraging us to develop this benchmark for platform performance architectures projections.

References

1. Dean, J., Ghemawat, S.: MapReduce: simplified data processing on large clusters. In: OSDI (2004)
2. Chen, Y., Asplaugh, S., Katz, R.: Interactive analytical processing in big data systems: a cross industry study of MapReduce workloads. In: International Conference on Very Large Data Bases (VLDB), Aug 2012
3. Chen, Y., Ganapathi, Griffith, R., Katz, R.; The case for evaluating MapReduce performance using workload suites. In: 19th IEEE International Symposium on Modeling, Analysis and Simulation of Computer and Telecommunication Systems (MASCOTS) (2011)
4. Baru, C., Bhandarkar, M., Nambiar, R., Poess, M., Rabl, T.: Benchmarking Big Data systems and the BigData Top100 list. Big Data (IMPETUS Innov. Archit.) **1**(1), 60–64 (2013)
5. Huang, S., Huang, J., Dai, J., Xie, T., Huang, B., The HiBench benchmark suite: characterization of the MapReduce-based data analysis. In: ICDEW (2010)
6. Wiki, PigMix Benchmark. http://wiki.apache.org/pig/PigMix
7. GridMix3 – Emulating Production Workload for Apache Hadoop. https://git.apache.org/hadoop-mapreduce.git/src/contrib/gridmix
8. STAC: Comparison of IBM Platform Symphony and Apache Hadoop Using Berkeley SWIM. STAC, LLC. Nov 2012
9. SWIMProjectUCB, 2012. https://github.com/SWIMProjectUCB/SWIM/wiki
10. Jia, Y., Shao, Z.: A Benchmark for Hive, Pig, and Hadoop. https://issues.apache.org/jira/browse/hive-396
11. Thusoo, A., Sen-Sarma, J., Jain, N., Shao, Z., Chakka, P., Anthony, S., Liu, H., Wyckoff, P., Murthy, R.: Hive - a warehouse solution over a Map-Reduce framework. In: VLDB (2009)
12. Rumen. http://hadoop.apache.org/docs/r1.1.2/rumen.html
13. Zujie, R., Xu, X., Wan, J., Shi, W., Zhou, M.: Workload Characterization on a Production Hadoop Cluster: A Case Study on Taobao. In: IISWC (2012)
14. TPC-C Benchmark. http://www.tpc.org
15. Poess, M., Floyd, C.: New TPC benchmarks for decision support and Web commerce. In: SIGMOD (2000)
16. The HiBench Suite. https://github.com/intel-hadoop/HiBench
17. Wikipedia. Gamma Distribution. http://en.wikipedia.org/wiki/Gamma_distribution
18. Krishnan, K., Saletore, V.A.: Sysviz: system visualizer for cluster performance characterization. Internal report, Intel. Corp (2012)

Applications and Scenarios

Big Data Workloads Drawn from Real-Time Analytics Scenarios Across Three Deployed Solutions

Tao Zhong$^{(\boxtimes)}$, Kshitij Doshi, Xi Tang, Ting Lou, Zhongyan Lu,
and Hong Li

Software and Services Group, Intel, 20th floor, Tower D,
Beijing Global Trade Center, Beijing, China
{tao.t.zhong,kshitij.a.doshi,xi.tang,ting.lou,
zhongyan.lu,hong.li}@intel.com

Abstract. Big Data solution vendors and customers alike face a pressing need for a few credible benchmarking workloads for demonstrating or optimizing performance, elasticity, efficiency, and robustness of solutions they create or deploy. Many new problems require extraction of immediately actionable intelligence from torrents of data, so a good application level benchmark must reflect in its design both real-time (low latency) and high throughput metrics. It should also impose loads that reflect the realities of complex, interdependent mixes of storage and analysis operations. This short paper describes three different application level scenarios. In these scenarios Big Data solutions are used to generate answers in real time for a subset of requests while requests that do not require such real time responses are completed at high rate in the background in presence of massive inflows of new data. The solutions from which we draw these scenarios are already in deployment or in pre-deployment testing, and thus can serve as good models from which to draw design perspectives in assembling a realistic Big Data workload, meaningful to customers tackling real-time needs while balancing high availability and service rate requirements.

Keywords: Real-time analytics · Data processing · Performance · Latency · Transactions · Workload · Benchmark · Databases

1 Introduction

Institutions and businesses in recent times have raced to capitalize quickly upon information their operations generate. High velocity inflows of data have made it difficult to deploy traditional relational database management systems (RDBMS) whose ETL requirements and transactional consistency overheads can impede rapid extraction of value from unstructured data. At the same time, most enterprises rely to a great degree on the robustness and strict transactional guarantees of RDBMS solutions for their business critical processes. This tension has spurred an enormous amount of innovation in recent years, with new options for processing of structured, unstructured, and semi-structured data emerging from hundreds of solution providers. Alongside this rapid innovation has emerged a critical need for credible benchmarking workloads that

© Springer International Publishing Switzerland 2014
T. Rabl et al. (Eds.): WBDB 2013, LNCS 8585, pp. 97–104, 2014.
DOI: 10.1007/978-3-319-10596-3_8

can represent in a meaningful way the stresses and the figures of merit by which these solutions can be analyzed and compared with each other.

In response, the scientific, industrial, and academic communities have come together to weigh the wide ranging criteria and define a standard Big Data benchmark [1]. This short paper contributes to this effort by drawing workload examples from three real-life solutions in which new information is instantaneously integrated into queried datasets. The paper offers examples in which answers are generated in a few seconds for a subset of requests, while requests that do not require such real time responses are completed at high rate. In practice, the sizes of datasets that a solution must handle in a resource efficient manner and the velocity with which it must process requests are both vital measures of its effectiveness, but alongside the blend of latency sensitive and throughput sensitive computations, different degrees of freedom exist for each. Specifically, there are situations in which a fast answer is needed within a matter of seconds but where the data to be examined is generally limited in size or the question to be answered is simple (e.g., "Did X occur in last Y intervals?"); and conversely where data to be examined is potentially vast or the question to be answered requires significant amount of computation, it is common to receive an answer in minutes (e.g., Multivariate analysis). We draw these observations from our work with medium to large scale corporations in China in creating such blended solutions [12, 13]. The testing of those solutions before and during deployment thus gives us real-life reference points and nuances to consider in assembling a Big Data workload.

The paper is organized as follows. Section 2 sets out the motivation, explaining the need for representative big data workloads. Sections 3, 4, and 5 describe respectively the following workflows and their characteristics: *Smart City*, *Content Management*, and *Fraud Prevention*. Section 6 summarizes the paper.

2 Motivation and Related Work

Big Data describes data sets so large and so complex that one must disperse data processing across many machines using distribution friendly approaches to solve problems in a timely, efficient, and resilient manner. The solutions vary considerably from one another owing to competing design objectives, frameworks, and usage scenarios; for example, a scale-out cluster like HBase [2] can boost throughput by adding more machines rapidly, but its batch processing latencies can make it difficult to employ for low-latency operational analytics; or an in-memory relational database can support fast query processing but may require costly and cumbersome disaster recovery provisions. Customers of Big Data solutions thus need a common benchmarking workload to gauge the relative merits of different solutions, from either hardware or software perspective.

Let us discuss briefly the more popular among the many tests of performance presently available. Apache Hadoop [3] includes several workloads such as TeraSort [4], Wordcount [5], and others that can flush out hardware bottlenecks such as inadequate bisection bandwidth or insufficient storage throughputs in relation to the volume of CPU horsepower available in a cluster. Yahoo! Cloud Serving Benchmark [6] exercises several NoSQL [2, 7, 8] solutions using a simple data model and varying mixes of reads and writes, and can be readily extended to other NoSQL solutions. HiBench [5] contains a mix of 9 Hadoop based operations, which span a few micro-benchmarks (like TeraSort), HDFS tests, and searches. Other benchmarks such as SWIM [9] and GridMix [10] while not representing any common usage scenario stress a cluster under test with similar mixes of low level operations. Noting that one common use of Big Data approach is in extracting business insights, BigBench [11] proposes a system level approach for performance testing by augmenting the industry standard TPC-DS workload with semi-structured and unstructured data operations.

In practice, customers are embracing blended solutions – using traditional RDBMS for a subset of their business critical operations (e.g., processing of payments), while distributing others over hundreds of nodes using application managed consistency protocols [12]. Customers are also pursuing new solutions where dataset being queried includes data mutating as a result of ongoing operations [13]. In these solutions, measures of performance need to include latencies of such operations, alongside throughput and scalability of solutions. From a benchmarking standpoint it is thus necessary to include latency as a critical measure, and, to do so while reflecting the richness of solutions in which relational and non-relational operations share the data environment. In the following three sections, we will describe workflows from three solution that are being readied for deployment, that each contain the characteristic of blending real time with high throughput demands, and mixing relational with application managed data processing.

3   Smart City – Solution Flow, Workload, and Metrics

This section describes a solution in deployment testing for a city in China, characterized by a blend of:

- Structured and unstructured data processing
- Transactional and analytic activities
- Scale out in-memory processing combined with distributed persistent stores
- Real-time and batch operations
- Information inflows from sensor and non-sensor devices structured and unstructured data

The problem being solved is to identify violations of one or more traffic safety rules in real-time, particularly the use of illegal license plates to escape registration charges or identification. The workload consists of interactions between a real time analytics module distributed over a NoSQL cluster at the center of Fig. 1, and four other systems, as described below:

1. An extraction system (ES): Cameras from several thousand locations send time-stamped images to local servers where they are transformed into tuples describing license plate, color, timestamp, and other attributes of vehicles, and sent to the RTA cluster described next. The images themselves are tagged and saved in a distributed store.

2. A real-time analytics cluster (RTA): The RTA is a distributed in-memory service that identifies rapidly those instances in which the records it receives from ES point to potential inconsistencies —such as a vehicle that should be Green is Red or is suddenly too far from its previous known location. RTA notifies the enforcement system described in item 5 below when it detects inconsistencies. Further, after it has eliminated duplicates, and merged and summarized records for each vehicle, RTA uploads the data into a distributed file system (FS, described next).

3. File System (FS) cluster: The FS cluster buffers processed data from RTA, which can then re-reference it later. For example, if a vehicle is spotted in many far-away locations over multiple days then that can be a yellow flag for more extensive checking. FS also acts as a source of data for the RR component described next.

4. Registration Records (RR): RR holds relational data across multiple departments of motor vehicles, for static information binding license plates to vehicles and their descriptions, owners, and home locations. It also integrates summaries of behaviors uploaded from the FS cluster – such as average speeds, frequent locations, etc. associated with the vehicles.

5. Enforcement Engine (EE): Once RTA has found inconsistencies, it notifies an enforcement engine so that the nearest public safety department can take cognizance and act on its findings. EE may implement various policies and overrides it may consider useful; for example, it can prioritize urgent responses and dispatch the necessary personnel to investigate a suspicious vehicle, to assist a vehicle that may have stopped moving, and take other actions.

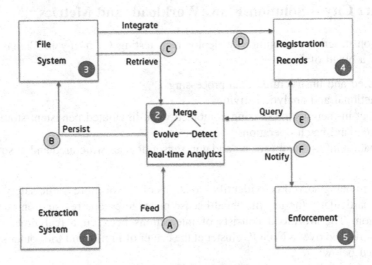

Fig. 1. Traffic safety solution

The traffic safety and compliance solution described above is being readied for operation, and represents a significant real world usage scenario in its information and processing flows–

1. Flows A, B, and C, as well as storage structures in FS and RTA, concern unstructured or semi structured data.
2. Flow D is an example of a background ETL operation. RR is structured as a traditional warehouse.
3. Flow F and the EE represent post-processing operations that may feed a complex event processing ensemble.
4. The processing at RTA is largely in-memory, since RTA must complete its work in real time for each uploaded record.
5. Flows B and C, together with the FS cluster, are an example of a large scale distributed object storage system.

We next describe test scenarios and metrics for the operations shown in Fig. 1. (1) The ingest rate for flow A: this is a primary measure of solution performance. For a given size of the total traffic stream, the RTA must operate without failing to identify suspicious records without exceeding a threshold number of seconds. The ingest rate needs to scale well with RTA capacity. (2) The object store and retrieve operations at the FS cluster need to scale in proportion with the ingest rate for flow A. (3) The data integration (ETL) rates at the data warehouse RR need to be sufficiently high so that querying operations (flow E) do not return stale information.

4   Content Management Under High Volume – Solution Architecture, Workload, Metrics

This section describes a Big Data application in deployment at a major content distributor that combines new and traditional media content into one collection for searching, filtering, editing, encoding and other operations. The distributor needs to index and cross-reference the content, while its volume keeps growing due to accumulation over time. The workflow blends

- Structured and unstructured data
- Transactional and analytic activities
- RDBMS and HBASE/Hive [16] operations.
- Searches over recent content finish in real time, while those over all content may take several minutes.

Figure 2 shows a block diagram for the solution. New data arrives through relational database operations. Data is bulk transferred from RDBMS to HBASE using Sqoop [17] at first, or at other times when major changes need to be transferred. Incremental changes to records in RDBMS are absorbed into HBASE through a log extraction and transformation module. Queries against the RDBMS complete quickly due to the limited volume of information. Both simple and complex queries over the full accumulation of data are performed against HBASE, either directly, or through Hive. Data analysis logic, typically originating in Java is bridged over to HBASE using object-relational mapping by HBase and Hive dialect modules. Many short (simple) queries benefit from in-memory accesses, since they generally touch hot records that tend to be in memory.

The interesting flows and their corresponding average latency and total throughput metrics from Fig. 2 are listed next: (1) the two methods of loading data into HBASE: (a) incremental and (b) via Sqoop, (2) Simple (primary key based) queries against data in HBASE via the HBASE OGM [18] driver, and (3) Complex Hive queries involving increasing numbers of attributes. How well the latency and throughput scale with cluster sizes under the constraint that cluster size must grow in tandem with dataset size, is another figure of merit.

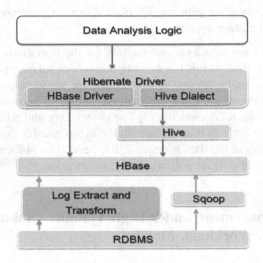

Fig. 2. Dealing with data accumulation over time

5   Electronic Fraud Detection and Prevention: A Simple Real-Time Workload

Compared with the previous two scenarios, the third Big Data usage scenario we describe is closer to a micro-workload. It is an instance in which a blend of structured and unstructured data processing is used to perform mid-transaction analytics in real-time. It concerns a particular type of fraud that affects billing and payments for telecommunications ("telecom") services. Customers in China use charging codes, – which can be obtained from third party credit systems, in order to add credit into their telecom service accounts. These codes are essentially flexible like currency notes – and thus anonymous, which opens the possibility of fraudulent charging through impersonation, phishing, or outright stealing of codes. A fraudster can escape detection by keeping the magnitude of such credit theft small and spread over many stolen codes, as victims may not be aware of small amounts of discrepancies even if they happen many times.

The solution employs a straight-forward flow. As shown in Fig. 3, recharge requests (❶) are redirected to an analytics module (❷) which applies a sequence of very fast heuristic checks before committing (❸) the recharge transactions. As a check must be done quickly, in-memory processing is done using a very thin table, called a staging table in which accumulated charge_time and charge_amount are maintained for each account, and the table is sharded across a cluster of machines. Analysis is performed over small groups of charge requests. An example of a simple heuristic is as below—

```
SELECT phone_num, SUM(charge_time), SUM(charge_amount) FROM
trans_tab
WHERE SUM(charge_time) > T1 AND SUM(charge_amount) > T2
GROUP BY phone_num ORDER BY SUM(charge_amount)
```

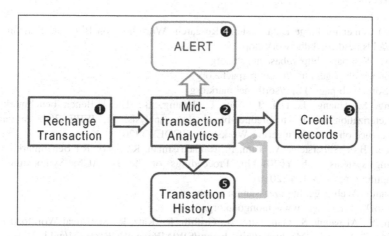

Fig. 3. Electronic fraud detection and prevention

which lets phone numbers with high recharge counts or amounts (exceeding thresholds T1 and T2) float to the top of the list for deeper inspection. Such transactions are sent forward to an engine which can perform more checks; other transactions can complete. Other more sophisticated tests can also be defined; i.e. variation from average call patterns, burst mode activities, and wide variations in charging may, for example, point to possible fraud. Latency of processing, both as a function of throughput and the size of the staging table, are the key metrics for evaluating this solution.

6 Summary

Across the three solutions and workloads used to test them, we have the following four themes in common- (a) they blend processing of structured and unstructured data; (b) they mix transactions with queries covering data that is recently updated; (c) they interweave short real-time queries with more complex queries which may execute outside real-time constraints; (d) they rely on in-memory processing for time critical answers. The smart city workload of Sect. 3 brings an interesting dimension of representing data inflows from sensor devices which one could emulate in a benchmark scenario by using just the extraction system outputs. The content management workload provides an opportunity to tie on-the-fly updates that arrive via an RDBMS into real-time queries that run against HBASE or Hive by using the log extraction logic. The fraud detection scenario can serve as a more contained micro-benchmark. Each usage represents a continuously engaged enterprise in which data affects and is affected by operational intelligence drawn from its rapid analysis. The characteristics identified above are likely to be increasingly common and therefore ideally represented in an industry standard Big Data benchmarking initiative.

References

1. UCSD Center for Large Data Systems Research, Workshops on Big Data Benchmarking. http://clds.ucsd.edu/bdbc/workshops
2. HBase Web page. http://hbase.apache.org
3. Hadoop Web page. http://hadoop.apache.org
4. TeraSort. Web page. http://sortbenchmark.org
5. Huang, S., Huang, J., Dai, J., Xie, T., Huang, B.: The HiBench benchmark suite: characterization of the mapreduce-based data analysis. In: IEEE 26th International Conference on Data Engineering Workshops (ICDEW) (2010)
6. Cooper, B.F., Silberstein, A., Tam, E., Ramakrishnan, R., Sears, R.: Benchmarking cloud serving systems with YCSB. In: Proceedings of the 1st ACM Symposium Cloud Computing, pp. 143–154 (2010)
7. Cassandra Web page. http://cassandra.apache.org
8. MongoDB Web page. www.mongodb.org
9. Chen, Y., Alspaugh, S., Ganapathi, A., Griffith, R., Katz, R.: Statistical Workload Injector for MapReduce (SWIM). https://github.com/SWIMProjectUCB/SWIM/wiki
10. GridMix Benchmark Web page. http://hadoop.apache.org/docs/r1.1.1/gridmix.html
11. Ghazal, A., Rabl, T., Hu, M., Raab, F., Poess, M., Crolotte, A., Jacobsen, H.: BigBench: towards an industry standard benchmark for big data analytics. In: Proceedings of the ACM SIGMOD, June 2013
12. Doshi, K.A., Zhong, T., Lu, Z., Tang, X., Lou, T., Deng, G.: Blending SQL and NewSQL approaches: reference architectures for enterprise big data challenges. In: IEEE Big Data Workshop, CyberC (2013)
13. Zhong, T., Doshi, K.A., Tang, X., Lou, T., Lu, Z., Li, H.: On mixing high-speed updates and in-memory queries: a big-data architecture for real-time analytics. In: IEEE BPOE Workshop (2013)
14. Spark Web page. http://spark.incubator.apache.org
15. Redis Web page. http://redis.io/
16. Hive Web page. http://hive.apache.org
17. Sqoop. http://sqoop.apache.org/
18. Hibernate OGM. http://www.hibernate.org/subprojects/ogm.html

Large-Scale Chinese Cross-Document Entity Disambiguation and Information Fusion

Xiaoge Li[✉], Sugang Ma, and Xiaohui Zhou

School of Computer Science, Xi'an University of Posts
and Telecommunications, Xi'an, China
{lixg,msg,xiaohuizhou}@xupt.edu.cn

Abstract. Cross-document entity disambiguation is the problem of identifying whether mentions from different documents refer to the same or distinct entities and rises in information fusion and automated knowledge base construction. In this paper, we describe a Chinese Information Extraction (IE) and fusion system based on Hadoop Framework, which involves document-level IE and corpus-level IE, a pipeline and multilevel modular approach to Name Entity Recognitions (EDR), entity relationship extraction and information fusion. In document-level IE, information associated with each mention of the name can be merged into rich profiles for entities based on our co-reference and alias modular, in corpus-level IE, entity disambiguation is performed based on agglomerative hierarchical clustering using Map Reduce. The visualized results of the entity centric information graph have been demonstrated.

Keywords: Information extraction · Information fusion · Natural language processing · Information network · Entity disambiguation · Map reduce

1 Introduction

Big data such large-scale heterogeneous data sources from different documents has raised various challenges in Information Extraction (IE) technology. Information fusion, a new emerging area derived from IE, aims to address these challenges. Information extraction is concerned with identifying information in unstructured documents. In most cases, this activity concerns processing human language texts by means of natural language processing. In the web, The information spread all over the internet from different source' the same entity can be referred to by more than one name string and the same name string can refer to more than one entity. Entity Disambiguation is key factor of information fusion.

Information fusion contains both document-level IE and corpus-level IE and involves various subtasks including Name Entity Recognitions (EDR) and entity relationship various subtasks including name entity recognitions and entity relationship extraction, coreference resolution within document and cross document entity disambiguation. Latest development of IE system (e.g. OpenNLP

© Springer International Publishing Switzerland 2014
T. Rabl et al. (Eds.): WBDB 2013, LNCS 8585, pp. 105–119, 2014.
DOI: 10.1007/978-3-319-10596-3_9

NameFinder[1], Illinois NER system[2] and StanfordNER system[3]) has made it possible to extract 'facts' (entities, relations and events), most IE systems focus on processing one document at a time, and except for coreference resolution, operate one sentence at a time. The output contains rich structures about entities, relations and events involving such entities. It is a significant step for IE to proceed from the document level to the corpus level. A central problem in corpus-level IE to determine if the same name in different articles refers to the same person. Normal approaches to this problem [3] derive features from the context surrounding the appearance of an entity in a document and then apply clustering algorithms that can group similar or related entities across all documents. A joint inference method based on cross-fact constraints in information network also has been applied to approach this goal [11]. These approaches aim for being exhaustive and grow exponentially in time with the increase in the number of documents. Recently researchers have studied methods to cross document entity Disambiguation as computing similarity between pairs of entity mentions on the usage of parallel and distributed architectures such as Apache Hadoop [4,5] to deal with such big data problem.

There are a few publications on studies of Chinese IE System and information fusion [6,7], Chen and Martin explore the Cross-Document Coreference Resolution (CDCR) task in both English and Chinese. Their work focuses on use of both local and document-level noun-phrases as features in their vector-space representation. In this paper, we describe a Chinese information extraction and fusion system based on Hadoop Framework, which involves both document-level IE and corpus-level IE, a pipeline and multi-level modular approach to EDR and entity relationship extraction, and introduces the concept of Entity Profile and information network. Finally we detail our approach of entity Profile disambiguation based on agglomerative hierarchical clustering and demonstrate the visualization results of information network with entity profile and its relations.

The remainder of this document is organized as follows. In Sect. 2, we introduce our Chinese information fusion system. In Sect. 3 we give a brief survey of our document-level IE engine and performance on NER and collocated entity relationships detection, and introduce entity profile concept. In Sect. 4. We detail our corpus level IE and focus on Chinese Person name entity disambiguation using agglomerative hierarchical clustering. In Sect. 5 we evaluate our methods with discussions. We also measured the impact of using external knowledge base such as Chinese Wikipedia, and present the visualization results of information network.

2   System Overview

The Information fusion was based on the foundation provided by our IE engine, which is a domain-independent, Pipeline and Hybrid Model architecture.

---

[1] http://opennlp.apache.org/
[2] http://cogcomp.cs.illinois.edu/demo/ner/?id=8
[3] http://www-nlp.stanford.edu/software/CRF-NER.shtml

Fig. 1 shows the overall system architecture involving the major modules. The dotted lines group the linguistic modules and IE modules together and the IE is supported by our linguistic modules.

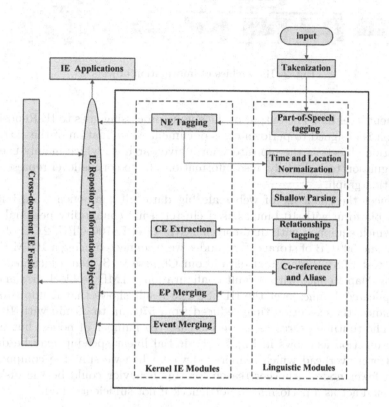

Fig. 1. Information fusion system architecture

This system distinguishes three layers: Chinese IE Engine, (cross-document) IE Fusion, and IE Applications. These three layers are linked by IE Repository, which stores the Information Objects such as Entity Profile and Events based on Hbase. The Hierarchies of Information Objects is shown in Fig. 2. The profile is an assembled information object for an entity. It was defined as an Attribute Value Matrix (AVM) to represent key aspects of information about entities, including their relationship with other entities. In our entity centric information network, each entity acts as the vertex and relations as edges linking each related entity profiles, forms as Information graph $G = G_i(V_i, E_i)$, where V_i is the collection of entity nodes, and E_i is the collection of edges linking one entity to the other, labeled by relation or event attributes, such as "relative_of" and "Place_of_birth".

For the IE Engine, a document is the largest unit for processing; this defines the scope for document-internal information merging based on discourse analysis

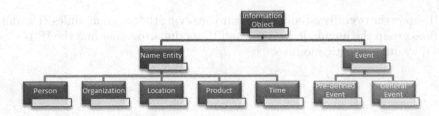

Fig. 2. Hierarchies of information objects

(Document level IE). The output of the engine processing goes to IE Repository. IE Fusion is designed to perform cross-document consolidation of the extracted information objects in the repository for a given archive, which mainly is entity disambiguation task. Finally, fuse information at the corpus-level represents as information graph.

To meet the challenge of web scale big data, all components are built in Hadoop platform with 10 Linux based cluster using commodity personal computer, which equipped with dual-core 3.2 GHz Intel E6700 CPU, 2 GB of main memory and 500 GB of storage. The nodes were connected using a 100 M Ethernet. We test 1.08 GB plain text data in our Chinese IE System (document level IE) using Map Reduce with default configuration: 64 MB as block size and the data replicated 1 times over the HDFS. The result shows that it improves the performance; the execution time reduced from 570 min to 39 min with 10 data nodes. The runtime decreases as we increase the number of nodes, but not in a linear function as shows in Fig. 3. The initial linear speedup may be due to the platform overhead which is more relevant when we split the computation between fewer nodes. Another reason for this behavior could be the disk I/O which could act as a performance bottleneck if not sufficiently high.

3   Document-Level Information Extraction

Information extraction on document level contains three tasks: named entities tagging, detecting relationships among named entities and entity profile merger.

At the engine layer, hybrid, multi-level pipeline architecture is used for extracting three major types of targeted information objects, namely, NE, EP and Event. The hybrid approach complements corpus-based machine learning by hand-coded Finite State Transducer (FST) rules. The essential argument for this strategy is that by combining machine learning methods [8–10] with an FST rule-based system [7], the system is able to exploit the best of both paradigms while overcoming their respective weaknesses. The system is modularized into distinct components according to the functions they perform to make each module a plug-and-play component of the system. This architecture serves as a reference framework for the research and development work for this effort.

As shown in Fig. 1, the core of the system consists of several kernel IE modules and linguistic modules. These modules remain domain independent. A striking

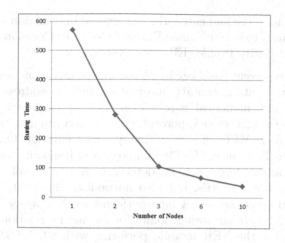

Fig. 3. Scalability to process for intra-document IE

feature of our system architecture is that the IE technology is built on multi-level NLP support, which includes:

- Chinese Word Segmentation (Tokenization) and POS tagging, which use Cascaded HMM method as described by Liu Qun [12].
- Shallow Parsing involves preliminary syntactic analysis. It matches patterns of simple, un-embedded linguistic structures and groups them into basic syntactic units. The major basic units are NPs, basic Adjective Phrase (AP), and basic Prepositional Phrases (PPs).
- Deep parsing decodes logical subject-verb-object and other grammatical relationships such as modification, conjunction and apposition. This involves a considerable amount of semantic analysis and provides a common structural basis for supporting high-level IE for relationships.
- Time Normalization interprets the time expression extract the temporal information with a cascade of finite automata [13] and represents it in ISO8601 format. Such as '2008 年 8 月 8 日晚上 8 点 8 分' will be Normalized as 20080 808200800. The location normalization is to identify the correct sense of a possibly ambiguous location Named Entity (NE)' such as '伦敦' (London) could be in England or in Canada, the methods applying for location Normalization in our system is similar to H. Li [13].
- Co-reference involves Discourse Analysis, dealing with structures that cross sentence boundaries. It consists of two sub-modules: (i) alias co-reference: e.g. '张明' with '小明', and '中国石油化工总公司' with '中国石化'; and (ii) anaphoric co-reference: decoding the NE referents for pronouns and anaphors, such as '他' for '李明' and '这个公司' for '中国电信'. Alias Co-reference is much more tractable than resolving anaphors. A hybrid model is being developed and benchmarked for anaphoric co-reference.

As shown in Fig. 1, the four categories of targeted information for our Chinese IE engine are defined to be: (i) Named Entities (NE), (ii) Correlated Entity (CE) relationships, (iii) Entity Profile (EP), (v) Events.

Named Entity represent basic key items such as proper names of person, organization, product, location, contact information such as address, email, phone number, URL, time, numerical expressions such as date, year, and various measurements like weight, money, percentage, etc., according to Message Understanding Conferences (MUC) of Named Entity Task Definition. Our system combined Finite State Transducer (FST)technology and Random Conditional Field (CRF) [15] for various Name entities detection. And it is not only tag time, location and measurement NEs, but also normalizes these entities to support entity Profile and event merger. A high-performance NE tagger is vital for EP extraction since NEs are anchoring points for extracting relationships. Manual checking shows that the NER module performs with 93.05–96.93 % accuracy (Table 1).

Table 1. Chinese NE benchmarking

	Person	Location	Organization	Time
Precision	94.05 %	96.93 %	92.77 %	92.95 %
Recall	93.11 %	91.66 %	82.98 %	90.67 %

Entity relationship captures correlated entity relationships within sentence boundaries In our system is implemented by the cascaded application of a list of pattern matching grammars. The results will be consolidated into the information object Entity Profile based on co-reference and alias support. Table 2 Shows the Benchmarking of some NE Person related Relationships.

Table 2. Personal relationships benchmarking

	Affiliation	Age	Birth place	Birth date	Relatives
Precision	98.53 %	96.67 %	100 %	100.00 %	100.00 %
Recall	71.28 %	76.32 %	70 %	83.78 %	78.57 %

Entity Profiles are complex rich information that collects entity-centric information, in particular, all the relationships that a given entity is involved in. This is achieved through document-internal fusion of related information based on support from co-reference and alias association. EP is an information object representing a real-world object named by a word string. Each defined relationship is represented by an attribute slot in the EP AVM. Each attribute-value pair gives some information about the entity in one aspect. The EP relationships consist of three types of information: (1) specific relationships such as

MPLOYEE_OF in MUC TR, (2) general relationships such as descriptors, modifiers and associated-entities, and (3) links to involved-events and entity profiles. Table 3 shows the example of Organization Profile. The progress from NE to Profile is believed to be a significant development in information modeling and IE representation. NE only represents one isolated aspect of information about an entity, i.e. its name. NE information alone is often not rich enough to facilitate the process of mapping an information object to the real world entity that it refers to. In comparison, the profile information object is a much richer structure of representation for an entity since key information about this entity is collected as a result of the IE process.

Table 3. Definition for organization EP

Attribute	Appropriate value	Comments
PrfID	Number	Profile ID
Type	Person|Organization| Location etc.	Pre-defined Profile type
Name	NE Person	Profile Name
Aliases	String	Alias and Co-reference chin
Date Of Birth	NE Time | NE Date	
Date Of Death	NE Time | NE Date	
Place of Birth	NE Location	Link to corresponding ProfileID
Position	String	eg. Chairman, CEO
AFFILIATION	NE Organization	Link to corresponding profileID
SPOUSE	NE Person	Link to corresponding profileID
PARENTS	NE Person	Link to corresponding profileID
CHILDREN	NE Person	Link to corresponding profileID
Age	NE age	
Quote	String	speech
ADDRESS	NE Address	
EMAIL	NE Email	
PHONE	NE Phone	
WWW	NE Website	
General Event	Verb String	
Pre-defined Event	String Event Name	Link to responding Event ID

There are two types of event, Pre-defined events and general event Pre-defined are defined by the user, e.g. events such as Earthquake, Company Acquisition, etc. General Events are verb-centric information objects representing 'who did what to whom when and where'. These events are domain independent.

The last step for document-level IE is Entity Profile (EP) Merging. EP merging is designed to merge multiple locally extracted relationships involving a given

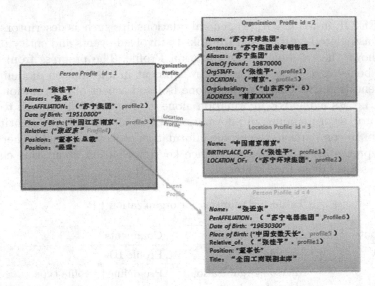

Fig. 4. Example of Profile linkage

entity into its discourse EP. This is accomplished with support from Co-reference and alias module. This module is also responsible for linking discourse EPs and their related events together. The results are richer and more condensed, as previously shown in the sample Person EP in Fig. 4. The results are discourse EPs output to the IE Repository on Hbase for cross document information fusion.

4   Corpus Level Information Extraction

It is a significant step for IE to proceed from document level to corpus level. During the course of our EP research, we found that the entity relationship information is by nature 'sparse data'. It is often the case that the majority of names in a document are mentioned with limited information about the entities to which these names refer. Only a few names are coupled with correlated information such as affiliation, position, descriptors, modifiers, associated-entities, etc. In such a situation, cross-document fusion of discourse EPs into corpus-level EPs is the key to enrich the information object to a useful level of content.

EP Fusion is a corpus-level module designed to further merge, consolidate and link EPs across documents in the repository. Merging enriches information, while consolidation involves eliminating redundant information, and linking connects this EP with other related EPs or events. In terms of EP extraction, this level is crucial to make the extracted information useful for information network construction.

A central problem in EP Fusion is cross-document entity co-reference, clustering of mentions into correct entities. For company and product EPs, merging can be based on string matching of the discourse EP's attribute name or aliases because their names (or brands) are often trademarked or uniquely registered. As to location EPs, merging can be performed supported by our location normalization. The key challenge for information fusion is to determine if the same name in different articles refers to the same person.

One of the most common approaches to cross document co-reference resolution has been based on agglomerative clustering, where vectors might be bag-of-word contexts [16,17]; Mann and Yarowsky [18] disambiguate person names using biographic facts, like birth year, occupation and affiliation. Artiles et al. [19] in the Web People Search Task clustered web pages for entity disambiguation. Recently entity disambiguation with a knowledge base becomes increasingly popular in the natural language processing community, several researcher have built a cross document coreference system using features from encyclopedic sources like Wikipedia, Freebase [20–22].

In our IE system, the person name entity is been detected in our Chinese IE engine, and the multiple locally extracted relationships involving a given entity are merged into its discourse EP. Our cross document entity disambiguation system relies on agglomerative hierarchical clustering algorithm [24] using richer and more condensed information of EPs and efficient calculation of similarity scores using map reduce.

As previously shown in Table 3, person entity profile contains: (1) Personal attributes; (2) NE-type Features; (3) BOW-type features.

Therefore, given two Person name profiles, we have implemented three similarity measures:

- Personal attributes: we have defined three different similarity levels based on corrected entity relationship sharing: sharing no relationships, relationship conflicts and consistence. Each level assigns different weight. The similarity will be

$$psim(P_1, P_2) = \sum_{commonce} w(ce_{1i} \cap ce_{2j}) \tag{1}$$

Here P_1 and P_2 is two Profiles, ce_{1i} and ce_{2j} represent ith relationship of P_1 and jth relationship of P_2 respectively.
- Named entity co-occurrence: not all kinds of NE co-occurrences are decisive. Sharing some common NEs such as "United States", "China", etc. may provide little information for name disambiguation. Log (d/df) was used to measure the information of a proper name.

$$rsim(P_1, P_2) = \sum_{i=1}^{n} log(\frac{d}{df_i}) \tag{2}$$

- VSM-based context similarity: the words within the same sentence of the person names are used to create a vector. The cosine measure between vectors was used as the similarity measure. For BOW-type feature, the similarity is computed by function

$$dsim(P_1, P_2) = \frac{\sum_{j=1}^{J} w_{1j} \times w_{2j}}{\sqrt{\sum_{i=1}^{M} w_{1i}^2}\sqrt{\sum_{i=1}^{N} w_{2i}^2}} \tag{3}$$

Where $w_{ij} = tf \times log\frac{D}{df}$, tf is the frequency of the term tj in the vector. D is the total number of documents. df is the number of documents in the collection that the term tj occurs in.

To combine these features for use with HAC, we consider simply concatenating individual feature vectors together to create a single feature vector, and compute cosine similarity as showing

$$prfsim(P_1, P_2) = \alpha * psim(P_1, P_2) + \beta * rsim(P_1, P_2) + \gamma * dsim(P_1, P_2) \quad (4)$$

If $prfsim \geq threshold$, the two person names is considered to be identical and two profiles will be merged into globe profile.

We also implement a Map Reduce algorithm [25] for computing pairwise profile similarity in large document collections. As shown Fig. 5, each of Profile has been stored on the Hbase, and similarity metrics has been distributional computed based on Map Reduce.

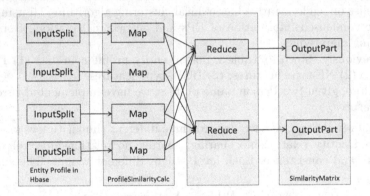

Fig. 5. Map Reduce for computing similarity metrics

We benchmarked our method using the standard purity and inverse purity clustering metrics as in the WePS evaluation. Let a set of clusters $S = \{S_1, S_2, \cdots\}$ denote the system's partition as aforementioned and a set of categories, $R = \{R_1, R_2, \cdots\}$ is the gold standard. The precision of a cluster S_i with respect to a category R_j is defined as,

$$Pur = \frac{\sum_{S_i \in S} max_{R_j \in R} |S_i \cap R_j|}{\sum_{S_i \in S} |S_i|} \quad (5)$$

And recall is defined as following

$$InvP = \frac{\sum_{R_i \in R} max_{S_j \in S} |R_i \cap S_j|}{\sum_{R_i \in R} |R_i|} \quad (6)$$

The F score is defined as in (7) is used in performance evaluation. $\alpha = 0.5$ is used to give more weight to inverse purity,

$$F = \frac{1}{\alpha\frac{1}{P} + (1-\alpha)\frac{1}{R}} \tag{7}$$

We present the clustering performances of the various methods in our system based on the different features of extracted profile. Each experiment uses HAC with average link method on hadoop. Since the number of clusters is not known, when to terminate the agglomeration process is a crucial point and significantly affects the quality of the clustering result. As shown in Fig. 6, We empirically determine the best similarity thresholds to be $= 0.28$ and $\alpha = 0.36$, $\beta = 1$, $\gamma = 0.6$ for the experiments on the web data prepared by Harbin Institute of Technology in China. The dataset contain 11876 documents and annotated with 50 Chinese person names.

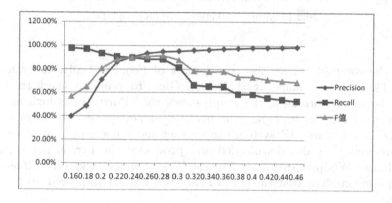

Fig. 6. CDCR results with threshold

Table 4. Testing results with different features

Features	Pur	InvP	F
PI	98.75 %	40.02 %	56.87 %
PI+NE	96.43 %	84.31 %	89.97 %
PI+NE+BOW	94.65 %	88.24 %	91.33 %

Table 4 shows the best results of our experiments on the training data sets with different feature configurations. We found that the personal information have far more discriminative power than most others in determining similarity between web pages, since person information, such as email address, and phone number is unique in EP. However it is not common such information can be extracted from Web page, the inverse purity is low with system running with person information only. The shared NE between EPs and BOW features are very helpful for this task in recall. The best F score have been achieved by combining all features for similarity measurement.

By combining all three features, we evaluate our methods using the benchmark test collection from Internet. We selected most common Chinese person name '李明' LI Ming and '王磊' WANG Lei and manually check the results. In the corpus there are 14407 and 6715 entity profiles respectively, totally more than 500 thousands person EPs, processed by our Document-level IE System. The testing results are shown in Table 5.

Table 5. Cross document coreference performance

Person name	Pur	InvP	F
王磊 WANG Lei	95.87 %	82.53 %	86.94 %
李明 LI Ming	95.06 %	80.68 %	84.25 %

For the comparison, we also test dataset of People's Daily, and randomly select Chinese person name '王刚' WANG Gang to manually check the result. We get better performance, the result shows that Purity is as high as 100 % and Inverse Purity is 99.25 %. It is because in newspaper, the form written language, our Chinese IE system can extract more corrected entity relationships precisely. We also downloaded and processed the person name entities from Chinese Wikipedia with our Chinese IE system as background knowledge. The experiment show that the performance has slightly improved up to 1 % in F score.

One of the initial goals for Information Extraction (IE) was to create a knowledge base from the entire input corpus, such as a profile or a event about any entity. In our system, the EPs will be merged and relationship among the EPs will be updated according to the results of Cross-Document Coreference Resolution. A heterogeneous information network has been constructed as defined in [26]. The information network represents an abstraction of the real world, focusing on the information object, as shown in Fig. 2, and the relationships between the objects, which is extracted by our IE system. A heterogeneous graph is denoted as $G = (V, E, \sum V, \sum E, l_V, l_E)$ where V is a set of vertices V Person EP, location EP, Organization EP, Pre-defined event,$\cdots$, etc; E is a set of directed edges (i.e., $E \subset V \times \sum l_E \times V$). In other words, we represent each edge as a triple $(v; l; v')$, which is a l-labeled edge from v to v', such as affiliation relationship between a person and an company.

Figure 7 shows our visualization result of a heterogeneous information network Blue nodes represent organization EPs, Organ are Location EP and grey nodes are Person entity profiles. When we select person to person network by filtering other information objects, it will be showing as the multi-dimensional network in Fig. 8. Person EPs connect each other with different types of relationship, such colleagues and relatives, etc.

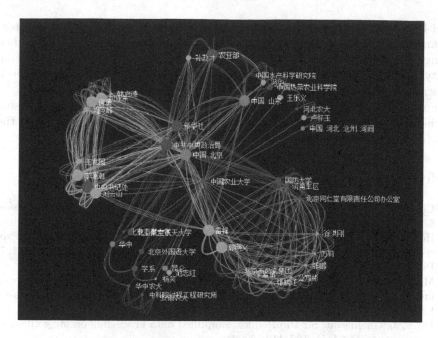

Fig. 7. Example of a heterogeneous information graph

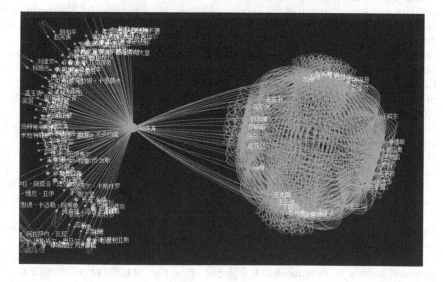

Fig. 8. Person EP multi-dimensional network shows person to person relationships

5   Conclusion and Further Work

In this paper we describe Chinese information fusion supported by the multi-level information extraction system based on Hadoop platform. In document level IE, the entity profiles have been extracted and merged. We present cross document name entity disambiguation based on agglomerative hierarchical clustering using rich information of entity profiles and BOW features to computing entity similarity. The entity centric heterogeneous information network has been constructed and visualization results haven demonstrated.

Future research is going on developing reliable and richer feature sets using machine learning technology [27] to improve both entity profile merger and the cross document coreference performance.

References

1. Chinchor, N., Marsh, E.: MUC-7 information extraction task definition (version 5.1). In: Proceedings of MUC-7 (1998)
2. Hobbs, J.R.: FASTUS: a system for extracting information from text. In: Proceedings of the DARPA Workshop on Human Language Technology, pp. 133–137. Princeton, NJ (1993)
3. Mayfield, J., Alexander, D., Dorr, B.J., et al.: Cross-document coreference resolution: a key technology for learning by reading. In: AAAI Spring Symposium: Learning by Reading and Learning to Read, pp. 65–70 (2009)
4. Dean, J., Ghemawat, S.: MapReduce: simplified data processing on large clusters. Commun. ACM 51(1), 107–113 (2008)
5. Singh, S., Subramanya, A., Pereira, F., McCallum, A.: Large-scale cross-document coreference using distributed inference and hierarchical models. In: Proceedings of the 49th Annual Meeting of the Association for Computational Linguistics, pp. 793–803, Portland, Oregon, June 19-24 2011
6. Ding, H., Xiao, T., Zhu, J.: A multi-stage clustering approach to chinese person name disambiguation. In: Proceeding of the 6th National Information Retrieval Conference, China (2010)
7. Silberztein, M.: Tutorial notes: finite state processing with INTEX. In: COLING-ACL'98, Montreal, Canada (1998)
8. Bikel, D.M., Schwartz, R.L., Weischedel, R.M.: An algorithm that learns what's in a name. Mach. Learn. 34(1–3), 211–231 (1999)
9. Borthwick, A.: Maximum Entropy Approach to Named Entity Recognition. Ph.D. thesis, New York University (1999)
10. McCallum, A., Li, W.: Early results for named entity recognition with conditional random fields, feature induction and web-enhanced lexicons. In: Proceedings of CoNLL, pp. 188–191, Canada (2003)
11. Li, Q., Anzaroot, S., Lin, W.-P., Li, X., Ji, H.: Joint Inference for Cross-document Information Extraction. In: CIKM' 11, Glasgow, Scotland, UK, October 2011
12. Liu, Q., Zhang, H.-P., Yu, H.-K., Cheng, X.-Q.: Chinese lexical analysis using cascaded hidden markov model. J. Comput. Res. Dev. 41(8), 1421–1429 (2004)
13. Chang, A.X., Manning, C.D.: SUTIME: a library for recognizing and normalizing time expressions. In: Proceedings of the Eighth International Conference on Language Resources and Evaluation (LREC-2012), Istanbul, Turkey, 23–25 May 2012
14. Li, H., Srihari, R.K., Niu, C., Li, W.: InfoXtract location normalization: a hybrid approach to geographic references in information extraction. In: Proceedings of NAACL-HLT Workshop on the Analysis of Geographic References, Edmonton, Alberta, Canada, 31 May 2003

15. Li, W., McCallum, A.: Rapid development of hindi named entity recognition using conditional random fields and feature induction. ACM Trans. Asian Lang. Inf. Process. **2**(3), 290–294 (2004)
16. Chen, Y., Martin, J.: Towards robust unsupervised personal name disambiguation. In: Empirical Methods in Natural Language Processing (EMNLP) (2007)
17. Bagga, A., Baldwin, B.: Entity based cross-document coreferencing using the vector space model. In: Conference on Computational Linguistics (COLING) (1998)
18. Mann, G.S., Yarowsky, D.: Unsupervised personal name disambiguation. In: Conference on Natural Language Learning (CONLL) (2003)
19. Artiles, J., Sekine, S., Gonzalo, J.: Web people search: results of the first evaluation and the plan for the second. In: WWW (2008)
20. Cucerzan, S.: Large-scale named entity disambiguation based on wikipedia data. In: Empirical Methods in Natural Language Processing (EMNLP) (2007)
21. Dredze, M., McNamee, P., Rao, D., Gerber, A., Finin, T.: Entity disambiguation for knowledge base population. In: Proceedings of the 23rd International Conference on Computational Linguistics, p. 285. Association for Computational Linguistics (2010)
22. Zheng, Z.C., Si, X., Li, F., Chang, E.Y., Zhu, X.: Proceedings of the 2012 IEEE/WIC/ACM International Joint Conferences on Web Intelligence and Intelligent Agent Technology, vol. 01, pp. 82–89 (2012)
23. Artiles, J., Gonzalo, J., Sekine, S.: The SemEval-2007 WePS evaluation: establishing a benchmark for the Web People Search Task. In: SemEval2007. ACL, June 2007
24. Jain, A.K., Narasimha Murty, M., Flynn, P.J.: Data clustering: a review. ACM Comput. Surv. **31**, 264–323 (1999)
25. Dean, J., Ghemawat, S.: MapReduce: simplied data processing on large clusters. Commun. ACM **51**(1), 107–113 (2008)
26. Lee, K., Liu, L.: Efficient data partitioning model for heterogeneous graphs in the cloud. In: The 13th International Conference for High Performance Computing, Networking, Storage and Analysis, Denver, CO, USA, 17–21 November (2013)
27. Downey, D., Etzionib, O., Soderland, S.: Analysis of a probabilistic model of redundancy in unsupervised information extraction. J. Artif. Intell. **174**(11), 726–748 (2010)

Tools, Systems, and Surveys

Big Data Operations: Basis for Benchmarking a Data Grid

Arcot Rajasekar$^{(\boxtimes)}$, Reagan Moore, Shu Huang, and Yufeng Xin

The University of North Carolina at Chapel Hill, Chapel Hill, NC, USA
{sekar, rwmoore, shuang, yxin}@renci.org

Abstract. Data Operations over the wide area network are very complex. The end-to-end implementations vary significantly in their efficiency, failure recovery and transactional management. Benchmarking for these operations is vital as we go forward given the exponential growth in data size. The critical evaluation of the types of data operations performed within large-scale data management systems and the comparison of the efficiency of the operations across implementations is an appropriate topic for benchmarking in a big data framework. In this paper, we identify the various operations that are important in large-scale data management and discuss a few of these in terms of data grid benchmarking. These operations form a set of core abstractions that can define interactions with big data systems by domain-centric scientific or business workflow applications. We chose these operational abstractions from our experience in dealing with large-scale distributed systems and with data-intensive computation.

Keywords: Benchmarking · Data grid · iRODS · Data operations · Optimization

1 Introduction

NSF has promoted a wide range of research and technology development projects as part of its Cyberinfrastructure Framework for 21st Century Science and Engineering (CIF21) [3]. These projects accelerate research, education and development of new functional capabilities in computational and data-intensive science and engineering with special emphasis on Data-Enabled Science and Community Research Networks. Projects funded by NSF, such as the iPlant Collaborative (plant biology data) [12], CUAHSI (hydrology data) [4], NEON (earth science data) [21], NEES (earthquake experimental data) [20], LSST (astronomy) [19], EarthScope (seismic data) [8], and OOI (marine data) [22] are assembling large amounts of data from sensors, simulations and analyses. Under the NSF DataNet program data life-cycle management projects such as DataONE (mainly earth science data) [5], DataNet Federation Consortium (multi-disciplinary data) [6], SEAD (multi-disciplinary data) [24], and TerraPop (mainly census data) [26] are integrating large-scale data collections with data sets from the long-tail of science. Aiming to provide community-engaged services and access to data for researchers and students, all of these projects provide access to data resources,

© Springer International Publishing Switzerland 2014
T. Rabl et al. (Eds.): WBDB 2013, LNCS 8585, pp. 123–137, 2014.
DOI: 10.1007/978-3-319-10596-3_10

metadata catalogs and remote services using a client-server architecture operating over the Internet.

The Big Data infrastructure is evolving towards a highly distributed system that manages multi-disciplinary data objects. A common set of principles drives the architecture of these data systems:

1. Federate distributed data for access by a community of authorized users through a portal,
2. Provide search, browse and discovery of data through discipline-centric catalog(s),
3. Control security and privacy through authentication and authorization,
4. Provide rich and flexible data processing in a service-oriented architecture,
5. Provide policy-based flexible data management to deal with heterogeneous requirements,
6. Maintain high availability and scalability through replication and distribution.

These six principles translate to a set of fundamental operations that are necessary for any type of data management or data provisioning system. The operations are implemented as application software modules in existing data management systems above an opaque Internet layer. Given the nature of highly distributed large data volumes, ignorance of networking can be a major bottleneck leading to common problems such as inefficiency, inflexibility, fragile security and costly software maintenance and upgrade.

In data grids, the intelligence needed to perform the operations is embedded in either the client and/or the server middleware. Protocols between clients and servers, and between peer-to-peer servers (in a federated system) are opaque messages sent between two physical IP-addresses. Operations such as discovery mechanisms for finding the location of a file from a logical address (e.g. DOI or logical collection name in iRODS and GFarm), access control mechanisms, authentication processes, and ingestion, placement and replication decisions are all done at the end points of the communication without taking advantage of any mechanisms that can be performed more efficiently at the communication level. Because of this mode of operation, one needs to perform many interactions between the end points, even to find the proper end points. These fundamental operations are independently designed and developed by each data management system. Hence, comparisons of approaches and benchmarking for these fundamental operations in data grid would be of interest for comparing capabilities of various implementations.

Database benchmarks such as TPC-C or TPC-DS [35], are based on broad categories of well-defined aspects of database applications. For example, "TPC-C simulates a complete computing environment where a population of users executes transactions against a database" and "is the new decision support benchmark that models several generally applicable aspects of a decision support system, including queries and data maintenance". These benchmarks are based on operations and transactions that have been observed and are representative of real world experience. Unlike databases and decision support system, the big data systems are still in their infancy and have not established clear and measurable sets of higher-level models of operations. This makes it difficult to define benchmarks for competing systems that track the performance of real-world systems. There is no single type of representative benchmark such as the order-entry transaction environment of TPC-C that can be defined at this stage.

Equivalent to transactions in databases, workflows form the basis for big data system operations. Workflows deal with large volumes of data that are possibly distributed and made of diverse varieties (formats and types) and a range of veracities (levels of integrity). Workflows can be viewed as distributed applications of varying complexity, running in parallel and concurrently with other workflows in a real life big data environment. Currently, workflows are defined in an ad hoc manner and for particular domains. For example, hydrology [28–30], astronomy [31], genetics [32, 33], climate [34] workflows have no common sequential core of operations that is amenable for defining a benchmark. Similar business-oriented workflows are also used in e-commerce and social-networking web applications. Hence, at this time defining workflow-based benchmarks for big data systems is premature and possibly not applicable in a wide context.

Our premise is that for the emerging big data systems useful and universally measurable (across systems) performance metrics need to be defined not at the application-level data-oriented operations, but at a lower middleware-level. This can be done by defining a cardinal set of essential operations that are widely used in a variety of workflows. This set of operations can then be used as a yardstick for comparison across implementations of big data systems over which the workflow applications are executed. There are two advantages in this approach. It gives a lowest common denominator for defining benchmarks which can be used to represent classes of problems independent of domain of applications, and it also gives a set of fundamental operations for characterizing scientific and business workflows in specific domains. Some examples of broad classes of generic problems that can be defined using the middleware-level data-oriented operations are:

- Caching and data placement strategies within a virtual network.
- Data transfer methods within a virtual collection.
- Policy enforcement strategies in and across virtual collections and virtual networks.
- Administrative task automation across both virtual networks and virtual collections.

In this paper, we identify multiple operations that are important in large-scale data management and discuss a few of these in terms of data grid benchmarking. In Sect. 2, we describe a data system that can be used as a model for defining the data-oriented operations. These operations are categorized and described in more details in Sect. 3. In Sect. 4, we discuss these operations in terms of benchmarking and operational measurements, and conclude the paper with Sect. 5.

2 Data-Oriented Systems

Research projects and academic institutions rely upon the ability to create and manage shared data collections to promote collaborative research. The collections are typically distributed across the participating sites, and are accessed through community-preferred mechanisms such as shell commands, web browsers, portals, and workflow systems.

Provisioning data services (such as reliable digital preservation, access, integration, and curation) through an interoperable data preservation and access network is central to Data-enabled Science. These provisioning services provide access to their distributed data collections and resources through web services, web portals, data grids, and data transfer protocols such as FTP and HTTP. Currently, the access to distributed data is mediated by access protocols (FTP, HTTP, etc.), distributed file systems (e.g. NFS, AFS [1], etc.), cloud storage systems (e.g. EC2 [9], SkyDrive [25], Dropbox [7], etc.), and data grids (e.g. iRODS [13], GFarm [11], the Globus Data Grid [10], etc.) which can be viewed as increasing levels of sophistication in delivering data between servers and clients.

A significant challenge exists in describing what constitutes a data-oriented system. The diversity of applications and plethora of significantly different system architectures makes it hard to define common terminology and common properties for a data system. One challenge is the development of a consistent namespace for describing the entities that are managed within a data management system, and for describing the operations performed upon the entities. Within the Research Data Alliance [23], three working groups are attempting to standardize operations: Data Citation working group, Data Foundation and Terminology working group and Practical Policy working group. The Data Citation working group seeks to provide unique identifiers to data subsets by associating a unique identifier for an operation with a unique identifier for the original data set. The combination then provides a unique identifier for the data subset. In a distributed environment, the combined identifier could be dynamically resolved into a remote operation that is performed upon a remote data set.

The Data Foundation and Terminology working group is defining the basic concepts required for data discovery and access, compared to those needed for managing a distributed data collection. If a workflow is registered into the data management system, then accessing the workflow should result in execution of the workflow and access to the workflow results. This implies that the entities managed within the environment are not just static data sets, but also include dynamic execution of workflows within a distributed environment. From this perspective, one can view a query on a database as an operation that is being performed at a remote site, with associated optimization of data that are returned.

The Practical Policy working group developed a concept graph describing policy-based data management. The concepts mapped from the driving purpose for a collection, to the properties that the collection should preserve. The properties identify assertions that the creators of the collection are willing to make about the collection contents. The properties are enforced by policies that control when and where procedures are applied to do the associated data management. The procedures generate state information that tracks the outcome of each procedure. The data management procedures comprise the basic workflows that enable automation of data retrieval and management of a collaboration environment.

We discuss the integrated Rule Oriented Data System in more detail, as part of our related work, because it has a superset of operations with respect to the other systems. Moreover, we have considerable experience in iRODS, and will be using it as a template for identifying data-centric operations in our research.

2.1   iRODS

The integrated Rule Oriented Data System [13–18] is developed by the Data Intensive Cyber Environments Center at the University of North Carolina at Chapel Hill and the University of California, San Diego. The iRODS Data Grid (Fig. 1) can be viewed as a network of fully-connected peer-nodes of resource servers, called iRES, which provide access to data and computational resources. The servers perform the protocol interchange needed for interfacing with exotic storage devices, and mapping them onto a uniform API used by the client, all built over the TCP/IP communication framework. The iCAT server holds the metadata used by the iRODS system and acts as a persistent store for the system status. The messaging server (iXMS) provides the means for the different servers (and services running in them) to communicate. In this way, services can be distributed, run in parallel, and communicate over time and space. The Scheduling Server (iSEC) allows the system to schedule jobs at a specific time, or periodically, or when a resource is available. iRODS federates distributed and heterogeneous data into a single logical file system (called the collection hierarchy) and provides a modular interface to integrate new client-side applications and server-side data and compute resources. iRODS also acts as a third-party mediator providing authentication, authorization and auditing, optimized data movement protocols and rich support for metadata at multiple levels of data collections. iRODS has a built-in distributed rule-engine. Administrators and collection owners can encode policies as rules for managing their data collections. These policies can be applied within the distributed server environment to realize operational functions such as:

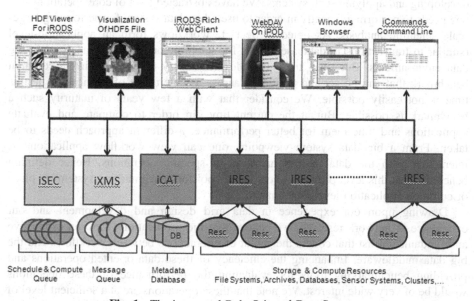

Fig. 1. The integrated Rule Oriented Data System

- *data accession workflows* (operations such as integrity and format/style checks, metadata extraction, pre-processing, relationship association with other data, replication, format translation that needs to be performed on ingestion of a new data object),
- *archival processes* (long-term management and assessment policies such as data migration, obsolescence control, provenance checking, periodic corruption checking, disposition enforcement, creation of vendor-independent Archival Information Packages),
- *publication and dissemination processes* (metadata association, indices for search and browse, generation of multiple formats for ease of access, distribution and replication for load balancing, access control checking, audit trails, notification),
- *analysis, synthesis and access processing* (operations such as discipline-specific analysis, integration of multiple data sets, capture of derived products with their provenance).

The rule engine in iRODS provides a way to customize the iRODS data grid to meet the demands of each discipline and also encode trust policies for sharing data across disciplines.

3   Operations in a Data Grid

The iRODS system provides an exemplar for discussing benchmarking in big data-oriented systems. Our familiarity with this system and its predecessor, the Storage Resource Broker [27], provides a unique view of the types of problems one can face in developing and applying such systems. We have abstracted a set of core operations that one needs to perform efficiently in order to use a big data system for performing large-scale scientific analysis. As noted in Sect. 1, workflows form an application-level (similar to transactions in databases) basis for scientific and business analyses in big data environments. But since these workflows are very diverse, finding a common benchmark that can be applied uniformly across a wide-range of applications at this time is not easily possible. We consider that with a few years of maturity such a benchmark is possible. But at the current time, in order to compare and evaluate applications and tune them for better performance, a different approach needs to be taken. From a big data system viewpoint, one can view workflow applications as interacting with the data middleware through specific operations. Hence defining benchmarks at this level provides a reasonable tool for benchmarking systems in place of common application-level benchmarks.

   Drawing upon our experience in data grid design and development and our experience in network research, we believe that these operations form an exhaustive and fundamental list that covers the gamut of interactions between workflows and the big data middleware. Enhancing the efficiency of these data-oriented operations and providing benchmarking tools for evaluating the performance of these operations would be of very wide interest. We note that these operations are at a sufficient level of granularity above low-level operations such as open file, read buffer, encrypt buffer, create directory, remove file, list files, etc. But they are still finer than the coarse application-level workflows. The operations in the list can be broadly categorized into four sets as follows:

A. Ingestion and Discovery:

1. Search & Discovery: Discover appropriate catalogs as well as search within catalogs,
2. Data and Metadata Ingestion: Ingest files into a disciplinary data grid as well as ingest metadata into a catalog; includes support for different formats of metadata,
3. Placement Strategies: File placement in multiple storage repositories based on user or administrator-defined policies,
4. Naming Support and Data Access: Access data within a disciplinary data grid, based on logical names (hierarchically-named content access) or universal identifiers (purl, uri, guid) and re-direct access to the appropriate storage site,

B. Data Movement:

5. Data transfer and backup: Optimize data movement (in bulk) between storage repositories,
6. Multi-copy support: Maintain multiple copies of a file (replicas) across multiple storage repositories and synchronize replicas across modifications and deletions,
7. Fault tolerance and failure recovery: Identify failure modes and switch to alternate service points for graceful degradation of data movement services,
8. Load balancing: Spread resource utilization to avoid hot points and bottlenecks,
9. Reservation and Quality of Service: Facilitate data transfer to meet user requirements,

C. Identification, Security & Integrity:

10. Authentication support: Authenticate users by directing to appropriate third-party authentication services (another aspect of discovery and re-direction)
11. Access Control support: Control access to resources (data, storage, compute, etc.) by directing to appropriate third-party authorization services (another aspect of discovery and re-direction)
12. Integrity Maintenance: Assist data collections by checking and maintaining integrity of their distributed replicas.

D. Data Manipulation:

13. Remote operations: Move the operation to the data, instead of moving the data to the operation (data subsetting, checksumming, etc.)
14. Distributed and parallel operations: Coordinate execution of operations across multiple storage locations (analyses at one location are integrated with analyses at another location).

A few of the above operations are good candidates for performance benchmarking similar to what we consider in database benchmarking. A few others are suitable for effective implementation checking. That is, even though they may not have execution performance issues, since they form an integral part of a workflow, checking for their achievability, effectiveness, and efficiency of implementation should be part of a benchmarking suite. A reason we need these two modes of benchmarking is due to the fact that there are no standards for big data systems and there are no exemplar workflow applications that one can set as a gold standard. So, not all operations are supported by all the data-oriented systems. As we advance in benchmarking these types of system, we will have a better handle at aggregating a definite set of benchmarks. Next, we discuss a few of these operations in more detail that are good candidates for performance benchmarks.

3.1   Search and Discovery

Search and discovery of datasets of interest is a fundamental operation. All scientific data repositories provide some means for identifying what is contained in each data file. This may range from a simple self-explanatory naming convention for the directory path and file name (e.g. /us/nc/ch/2011/feb/20/temp.txt) to a complex, queriable database containing metadata, possibly in a standardized schema. A data grid, like iRODS, provides a central place for domain-centric collections to register their data files in a common collection hierarchy. Finding the appropriate data catalog can by itself cause multiple problems because one needs to know the exact address of the location of the catalog. In the case of iRODS, each disciplinary data grid has an associated unique zone name. One can direct a query to a particular data grid (say hydrology) by connecting to any federated data grid and using the zone name as a parameter in the query:

imeta –z hydrology qu –d value_type = rainfall county = orange

In the above command, the hydrology metadata catalog is queried for all datasets who store values of type rainfall for Orange county. The query works even when one is connected to a different data grid (say marine) since iRODS provides a federation mechanism where each data grid catalog knows the location of other federated catalogs. A problem that is faced in such a federation is that as the number of data grids grows, as is the case for big data federations, the federation mechanism becomes unwieldy (n^2 federations are needed) or one needs to have a star-type federation with a master catalog. But this requires connection to the central node to perform querying which can lead to single point of failure. An appropriate benchmark would be to evaluate the performance of this metadata catalog implementation in checking whether the catalog querying operation is done effectively at the network level and how transparency is handled with changes in the locations and numbers of relevant metadata catalogs.

3.2   Data Movement

Data movement represents a class of operations that is central to performing replication, backup, caching or archiving from one server to another. Currently, the operations are agnostic of the underlying communication mechanisms. However, for future data fabrics, we believe that data and network resources have to be considered together as a whole so that optimal data transfer paths can be found. Performance benchmarking of data movement with awareness of the underlying network topology would be of wide interest. To illustrate the advantages, we list some candidate operational methods that will benefit from an architecture that integrates data management and data movement. Many of these operations are also good candidates for benchmarking initiatives:

1. Moving data in bulk between repositories. Broadly speaking, there are two types of communication mechanisms: connectionless and connection-oriented. In future data fabrics, the communication should be data oriented. The network should decide whether to use connectionless or connection-oriented communication based on the nature of the data and the type of operation. Suppose we need to migrate data in bulk from one site to another. One needs to look up a catalog first to figure out where the data are physically located. The result is passed to a Data Management system to decide, through other state information, the right storage systems to place the data (possibly with access control enforcement). Finally the transport controller needs to pick a specific communication mechanism (TCP, UDP, parallel or serial, and the number of parallel threads). Then the transfer takes place. Mechanisms to recover from error are also of interest as well as mechanisms to perform delayed data movement and provision support for quality of service. Data movement is fundamental for big data oriented systems. Performance for handling movement of a large number of small data files, and a small number of large data files, and a variable mix of them would be of interest to different user communities.

2. Multi-copy support is a typical operation in Content Delivery Networks (CDNs) where the data are cached in multiple places to facilitate faster user access. Again, this problem requires the joint management of data, storage and networking. Benefits of multi-copy support are realized as speedier access to 'popular' data, fault tolerance and load balancing. Benchmarking for performance with load variation and graceful degradation would be of interest to many scientific and business communities.

3. Fault tolerance and recovery. Computer equipment is subject to failures. To improve the reliability of data, it is possible to use network equipment to detect whether an end point is out of service (possibly due to server or network failures). Benchmarking mechanisms for identification of failure modes and recovery thereof would also be of interest. Even though failure detection is normally not considered a benchmarking goal, when dealing with a large number of systems (order of 10,000 s), the reliability of big data systems to detect these failures and find alternatives for recovering and resuming service, with minimal disturbance to the applications is of interest. Identifying the various modes of failure and developing performance metrics for evaluating recovery from these failures will be important to have in a big data benchmarking toolkit.

3.3   Access Control

Access of data from the data grid involves authorization. When a user accesses a file, one needs to check whether the user is indeed allowed to access the file by consulting an Access Control List or checking against a community database or service such as LDAP. In a data grid, these authorization services are either provided by the data grid's metadata catalog (iRODS and GFarm) or by a third-party service (Globus, which uses a CAS-based authorization [2]). In the case of a network-centric metric, this step can be efficiently done in two ways. In the first method, the router which sends the "get" command to a resource, can route the command such that it first goes to the authorization server which in turn "signs" the command and "forwards" the get command to the resource server. This requires that the router knows the location(s) of the authorization server for the data grid. This method of authorization can take advantage of a known catalog location system to find the location of the authorization service.

Evaluation of both authorization and authentication operations in a big data grid framework will be of interest, along with scalability for a large number of accessed objects and large user base, would be of interest. Performance issues crop up based on the types of authorization model that are chosen and in the way they are implemented. How systems perform (in terms of speed) when dealing with authorization of a large number of objects, or when dealing with a large number of concurrent requests, or when the size of the data collection is very large, need to be considered as benchmarking targets. Moreover, these authorizations may be inserted, updated and deleted and the behavior of the system when performing these operations in bulk, in parallel and in a distributed environment is of interest for benchmarking.

Load Balancing. Benchmarking of load balancing and parallel file transfer operations are of interest when considering data movement. Suppose data are replicated at different locations. It is possible to satisfy user data requests in a round-robin fashion or through randomization or through graduated learning strategies for load balancing. Performance testing is appropriate for whether autonomous server management can detect server activity dynamically to leverage load-balanced access. Parallel file transfers also leverage the fact that data may be replicated at various places. Somewhat resembling the P2P approach, a user may receive different parts of the data from different places in parallel. This approach in essence moves the bottleneck of data transfer from the source to the receiver and possibly onto the network. It has been observed that parallel data transfer is not a panacea and can actually degrade performance in some circumstances. Implementations (from greedy algorithms, to static algorithms to learning algorithms) can be varied in different big data systems with resulting performance issues. In a multigateway networks, parallelism can lead to bottlenecks and throttling may need to be done effectively to optimize performance. Benchmarking for performance is particularly interesting because it tests that data can be bifurcated and transferred separately to achieve better load balancing in a distributed system network.

Similar to the four operations considered above other data-oriented operations can be identified for benchmarking for performance and achievability.

4   Benchmarking

The many types of data management systems and workflow applications using them represent a distinct benchmarking challenge because it is not possible to find a single application profile that can cut across multiple domain-centric applications. Each data management application has a preferred class of data objects with associated properties.

These classes include small collections of large files, large collections of small files, distributed but virtually aggregated files, sensor streams with long-running data movement, short and bursty data transfers (ala twitter) with multi-cast needs, just in time video streaming with requirements on a large scale broadcast mode, etc. Even with these heterogeneous classes of data, one can easily see that they utilize the categories of operations that we have enumerated earlier. Data Operations over the wide area network are very complex and end-to-end implementations vary significantly in their efficiency, failure recovery and transactional management. Benchmarking for these operations is vital as we go forward given the exponential growth in data size. Moreover, data have a cumulative effect (compared to compute or networking functionalities). Namely, as the size of data increases, the behavior of the operational system can degrade in unexpected manners. Failure recovery in a wide area data management operational system, and the necessity to provide guarantee of service and atomicity for operations (i.e., the operations complete or leave the system in a stable state) also are important in big data management.

A rather common approach to evaluate a distributed system is to use micro-benchmarks on each individual building blocks. For example, for a cloud, different tools can be used to benchmark the components including CPU, disk I/O, memory bandwidth and networking separately. However, because most applications require multiple components working in concert, these micro-benchmarks only provide indirect results that can hardly be used to compare two different systems in a meaningful fashion. As a result, data-oriented macro-benchmarks are of interest.

As in software testing, treating the data system middleware as a black-box does not provide sufficient information to understand what causes performance degradation. To benchmark data grid systems, it is also necessary that within these systems, hooks be provided to the benchmarking suite to facilitate more detailed analysis. Specific to iRODS, benchmarking rules could be provided so that a data operation is decomposed into more basic actions if necessary. For example, a data movement operation can be roughly separated into two (maybe intertwined) actions: name resolution and network transmission. With an iRODS rule, the time spent on each action can be recorded individually for benchmarking or debugging purposes. While benchmarking the data movement operation is important to compare the performance of data grids, finer measurement granularity ensures one is not comparing apples to oranges.

4.1 Benchmarking Approach

In this section, we outline a possible approach that can be taken in defining and performing benchmarks on a big data-oriented system. As we stated before workflows are equivalent to transactions in databases and hence form ideal vehicles for testing and benchmarking. But most domain-centric workflows are very narrow and have many characteristics that are not relevant to other workflows in other domains. Hence, finding an exemplar set of workflows (as done with order-entry transactions in TPC-C) is not easy and possible at this time. Hence, we propose that a set of "synthetic" workflows be developed with community input. These synthetic workflows can be used to test small but well-defined sequences of operations that perform tasks that can be agreed by many communities of practice. These workflow modules should have relevant and clearly understandable goals with well-defined measurable and repeatable metrics that can be easily translated to discuss performance across a large number of domains and data oriented systems. These workflows can be drawn upon to provide broad acceptability, with scalable and applicable performance across a spectrum of hardware and middleware architectures running typical applications.

As is necessary when defining a suite of synthetic workflows, one also needs to define a set of good data collections on which these workflows are executed. Selection of these datasets needs to be done with care and with input from multiple domains. Since the data are of widely different sizes and types, care should be taken to find data that are average and measure operational performance in commonly implemented systems. As in TPC-C (which uses a specific schema of 9 tables) finding a single set of data collections would be ideal. But, we believe that big data systems may not have such a single set and may need to consider multiple sets of data; hopefully a very small number. Indeed, it might be necessary to define exemplar data collections for each domain of application since the types and formats of data used in climate or hydrology differ significantly from those used in genomics or neuro-informatics. Apart from varieties of data, one needs to consider other characteristics of big data – the 3Vs (Volume, Velocity, Veracity) – when finding appropriate data collections. Again, we consider that input from the communities of practice will be necessary to identify data collections of interest.

A third variable for consideration in big data benchmarking is the architecture under which the system will be tested. The reason that this is important in the case of big data is that there are several kinds of architectures that come into play in big data systems. Single site systems, to multi-site systems, to loosely federated but administratively independent systems, to third-party cloud systems to web-service based systems are some of the broad classes of systems that can be identified. Various topologies that can do the interconnections also provide another dimension to be considered. The workflow benchmarks may behave differently in these systems (or it may be proved that architectures do not matter!) and hence applications of benchmarks to commonly useable architectures may need to be identified.

Synthetic workflows, data collections and reference architectures provide a basis for developing benchmark for big data-oriented system based on middleware-level operations. In the rest of this section, we discuss a "strawman" workflow that can be a candidate synthetic workflow in the benchmark suite. Our workflow is based on our experience and caters to the needs of data managers who are building large-scale preservation systems which can be domain agnostic. In such a system the data managers want to guarantee conformance of seminal archive properties such as integrity, authenticity, chain of custody, and original arrangement.

Figure 2 lists the concepts that are involved in building such an archival data management system. Policy-based systems implement management policies as computer actionable rules and implement data management workflows as computer executable workflows. The computer actionable rules are stored in a rule base, and executed through a distributed rule engine. The fundamental operations performed in the distributed system correspond to well-defined workflows, and can be considered as candidates for benchmark workflows. Interactions with the system are done through policy-enforcement-points that allow the rule engine to choose the policies that should

Policy-based Data Management – Implementation in iRODS

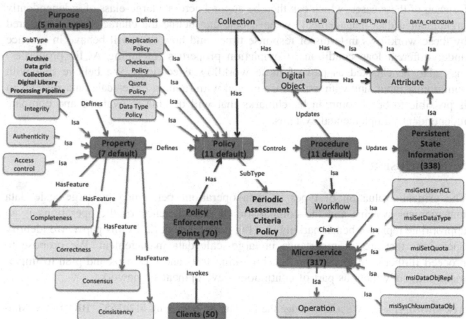

Fig. 2. Policy-based data management concepts

be enforced, and then execute the related procedures. Verification of properties is done through policies that are run periodically. Each policy is implemented through chaining of multiple operations. For integrity verification, the location of the replica must be found, a remote checksumming operation is applied, the result is compared with the saved value, and a report is generated. Each operation generates state information that must be saved. The state information may be queried to monitor status in a distributed environment, and can be audited to verify point in time assertions. Every archival data management system implements some form of these concepts. The implementation may hard code the properties and policies in software, or may rely upon an external service. A reduced set of operations may be supported, such as POSIX I/O, and the operations may be restricted to a local system. The generic system performs the operations in a distributed environment, in which data manipulation and network communication are closely interleaved.

In terms of synthetic workflows, one can define several procedures based on the Fig. 2 for archival data management systems. Just to provide a point of reference, to ensure integrity, multiple workflows are needed:

1. Manage a checksum for each file for verifying whether the file has changed.
2. Create replicas of each file.
3. Distribute replicas to different storage systems.
4. Periodically validate the existence of the replicas and their integrity.

Synthetic workflows can be developed based on the operations defined in Sect. 3 for each of these cases, which can then be applied across a large-class of independently implemented data management systems. One can define the characteristics measured by these workflows in terms of response times and how they will behave in practice under different load conditions. Completion properties (including ACID properties) may also be checked as part of these workflow executions. We believe that with community input, and with experience gained with running large-scale data systems, it is possible to build common benchmarks that will be able to evaluate and compare independently implemented systems.

5 Conclusion

The critical evaluation of types of data operations performed by large-scale data management systems and the comparison of the efficiency of the operations is an appropriate topic for benchmarking in a big data framework. In this paper, we identify operations that are of importance in large-scale data management. We propose to proceed further with identifying benchmarking tests and scenarios and plan to implement and study them as part of continuous development framework.

Acknowledgement. We acknowledge the funding by NSF grant #1247652 "BIGDATA: Mid-Scale: ESCE: DCM: Collaborative Research: DataBridge - A Sociometric System for Long tail Science Data Collections", by NSF grant #0940841 "DataNet Federation Consortium" and by NSF grant #1032732 "SDCI Data Improvement: Improvement and Sustainability of iRODS Data Grid Software for Multi-Disciplinary Community Driven Application".

References

1. OpenAFS. http://www.openafs.org/
2. Community Authorization Service. http://toolkit.globus.org/toolkit/docs/4.0/security/cas/
3. NSF: Cyberinfrastructure Framework for 21st Century Science and Engineering (CIF21). http://www.nsf.gov/about/budget/fy2012/pdf/40_fy2012.pdf
4. CUAHSI: Consortium of Universities for the Advancement of Hydrologic Science, Inc. http://www.cuahsi.org/his.html
5. DataONE: Data Observation Network for Earth. http://www.dataone.org/
6. DFC: The Datanet Federation Consortium. http://datafed.org/
7. The DropBox. https://www.dropbox.com/
8. EarthScope: Exploring the Structure and Evolution of the North American Continent. http://www.earthscope.org/
9. Amazon Elastic Compute Cloud. http://aws.amazon.com/ec2/
10. The Globus Data Grid Effort. http://www.globus.org/toolkit/docs/2.4/datagrid/
11. The Gfarm File System. http://datafarm.apgrid.org/
12. The iPlant Collaborative. http://www.iplantcollaborative.org/
13. iRODS: Data Grids, Digital Libraries, Persistent Archives, and Real-time Data Systems. https://www.irods.org

14. Moore, R., Rajasekar, A.: Rule-based distributed data management grid. In: 2007 IEEE/ACM International Conference on Grid Computing (2007)
15. Moore, R., Rajasekar, A., de Torcy, A.: Policy-based digital library management. In: International Conference on Digital Libraries, Delhi, India, 24–26 February 2009
16. Rajasekar, A., Wan, M., Moore, M., Schroeder, W.: A prototype rule-based distributed data management system. In: HPDC Workshop on Next Generation Distributed Data Management, Paris, France (2006)
17. Rajasekar, A., Moore, R., Wan, M., Schroeder, W., Hasan, A.: Applying rules as policies for large-scale data sharing. In: 1st International Conference on Intelligent Systems, Modelling and Simulation, Liverpool, UK, 27–29 January 2010
18. Wan, M., Moore, R., Rajasekar, A.: Integration of cloud storage with data grids. In: The Third International Conference on the Virtual Computing Initiative, Research Triangle Park, NC, 22–23 October 2009
19. LSST: The Large Synoptic Survey Telescope. http://www.lsst.org/lsst/science/development
20. Brown, G.E., Jr.: NEES: Network for Earthquake Engineering Simulation (NEES). http://nees.org
21. NEON: The National Ecological Observatory Network. http://www.neoninc.org/
22. OOI: The Ocean Observatory Initiative. http://www.oceanobservatories.org/
23. RDA: The Research Data Alliance. https://www.rd-alliance.org
24. SEAD: Sustainable Environment - Actionable Data. http://sead-data.net/
25. Microsoft SkyDrive. http://www.skydrive.com
26. TerraPopulus: Integrated Data on Population and Environment. http://www.terrapop.org
27. Baru, C., Moore, R., Rajasekar, A., Wan, M.: The SDSC storage resource broker. CASCON First Decade High Impact Papers, November 30–December 3 1998 (Reprint), pp. 189–200. doi:10.1145/1925805.1925816
28. Guru, S.M., Kearney, M., Fitch, P., Peters, C.: Challenges in using scientific workflow tools in the hydrology domain. In: 18th World IMACS/MODSIM Congress, Cairns, Australia, 13–17 July 2009. http://www.mssanz.org.au/modsim09/I8/guru.pdf
29. VIC: Variable Infiltration Capacity Macroscale Hydrologic Model. http://www.hydro.washington.edu/Lettenmaier/Models/VIC/
30. RHESSys, Regional Hydro-Ecologic Simulation System. http://fiesta.bren.ucsb.edu/~rhessys/index.html
31. Schaaff, A., Verdes-Montenegro, L., Ruiz, J.E., Santander Vela, J.: Scientific workflows in astronomy. In: Ballester, P., Egret, D., Lorente, N.P.F. (eds.) Proceedings of a Conference held at Marriott Rive Gauche Conference Center, Paris, France, 6–10 November 2011. ASP Conference Series, vol. 461, p. 875. Astronomical Society of the Pacific, San Francisco (2012)
32. Ghosh, S., Matsuoka, Y., Asai, Y., Hsin, K., Kitano, H.: Software for systems biology: from tools to integrated platforms. Nat. Rev. Genet. **12**, 821–832 (2011). doi:10.1038/nrg3096. http://www.nature.com/nrg/journal/v12/n12/full/nrg3096.html
33. Jimenez, R.C., Corpas, M.: Bioinformatics workflows and web services in systems biology made easy for experimentalists. Methods Mol Biol. (2013). doi:10.1007/978-1-62703-450-0_16. 1021:299-310. http://www.ncbi.nlm.nih.gov/pubmed/23715992
34. NARR: NCEP North American Regional Reanalysis. http://www.esrl.noaa.gov/psd/data/gridded/data.narr.html
35. TPC: Transaction Processing Performance Council. http://www.tpc.org/default.asp

BDGS: A Scalable Big Data Generator Suite in Big Data Benchmarking

Zijian Ming[1,2], Chunjie Luo[1], Wanling Gao[1,2], Rui Han[1,3], Qiang Yang[1],
Lei Wang[1], and Jianfeng Zhan[1(✉)]

[1] State Key Laboratory Computer Architecture, Institute of Computing Technology,
Chinese Academy of Sciences, Beijing, China
{mingzijian,luochunjie,gaowanling,yangqiang,zhanjianfeng}@ict.ac.cn,
harryandlina2011@gmail.com, wl@ncic.ac.cn
[2] University of Chinese Academy of Sciences, Beijing, China
[3] Department of Computing, Imperial College London, London, UK

Abstract. Data generation is a key issue in big data benchmarking that aims to generate application-specific data sets to meet the 4 V requirements of big data. Specifically, big data generators need to generate scalable data (Volume) of different types (Variety) under controllable generation rates (Velocity) while keeping the important characteristics of raw data (Veracity). This gives rise to various new challenges about how we design generators efficiently and successfully. To date, most existing techniques can only generate limited types of data and support specific big data systems such as Hadoop. Hence we develop a tool, called Big Data Generator Suite (BDGS), to efficiently generate scalable big data while employing data models derived from real data to preserve data veracity. The effectiveness of BDGS is demonstrated by developing six data generators covering three representative data types (structured, semi-structured and unstructured) and three data sources (text, graph, and table data).

Keywords: Big data · Benchmark · Data generator · Scalable · Veracity

1 Introduction

As internet becomes faster, more reliable and more ubiquitous, data explosion is an irresistible trend that data are generated faster than ever. To date, about 2.5 quintillion bytes of information is created per day [1]. IDC forecasts this speed of data generation will continue and it is expected to increase at an exponential level over the next decade. The above facts mean it becomes difficult to process such huge and complex data sets just using traditional data processing systems such as DBMS. The challenges of capturing, storing, indexing, searching, transferring, analyzing, and displaying *Big Data* bring fast development of big data systems [10,13,15,24,30]. Within this context, big data benchmarks are developed to address the issues of testing and comparing such systems, thus measuring their performance, energy efficiency, and cost effectiveness.

© Springer International Publishing Switzerland 2014
T. Rabl et al. (Eds.): WBDB 2013, LNCS 8585, pp. 138–154, 2014.
DOI: 10.1007/978-3-319-10596-3_11

Big data generation, which aims to generate application-specific data sets for benchmarking big data systems, has become one of the most important features of a benchmark. Generally, using real data sets as workload inputs can guarantee data veracity in benchmarking. Based on our experience, however, we noticed that in many real-world scenarios, obtaining real data sets for benchmarking big data systems is not trivial. First, many owners of real big data sets are not willing to share their data due to confidential issues. Therefore, it is difficult to get real data sets with a variety of types including structured, semi-structured or unstructured data. Moreover, it is difficult to flexibly adapt the volume and velocity of fixed-size real data sets to meet the requirements of different benchmarking scenarios. Finally, the transferring of big data sets via the internet, e.g. downloading TB or PB scale data, is very expensive.

A natural idea to solve these problems is to generate synthetic data used as inputs of workloads on the basis of real-world data. In [17], Jim Gray discussed how to efficiently generate data using pseudo-random sequence and a non-uniform distribution to test the performance of database systems. In recent years, with the rapid development of big data systems and related technologies, the 4 V properties of big data are required to be kept in the generation of synthetic data: (1) high volume and velocity data sets can be generated to support different benchmarking scenarios; (2) a diversity of data types and sources should be supported (variety); (3) the important characteristics of raw data should be preserved (veracity). To date, although some powerful data generators such as PDGF [26], a parallel data generator designed to produce structured data, have been proposed, a majority of data generators in current big data benchmarks only focus on specific data types or targets one specific platform such as Hadoop. Moreover, little previous work pays attention to keeping veracity of the real life data during data generation. In conclusion, current work has not adequately addressed issues relating to keeping the 4 V properties of big data.

In this paper, we introduce Big Data Generator Suite (*BDGS*), a comprehensive tool developed to generate synthetic big data while preserving 4 V properties. The data generators in BDGS are designed for a wide class of application domains (such as search engine, e-commence, and social network) in big data systems, and will be extended for other important domains. We demonstrate the effectiveness of BDGS by developing data generators based on six real data sets that cover three representative data types (structured, semi-structured, and unstructured data), as well as three typical data sources (text, graph, and table). In any data generator, users can specify their preferred data volume and velocity in data generation. Theoretically, in BGDS, the volume and velocity of a data set can only be bounded by the storage capacity, the hardware configuration (e.g. the number of nodes), the parallelism strategy (how many generators execute in parallel), and the total execution time. At present, BDGS is implemented as a component of our open-source big data benchmarking project, *BigDataBench* [14, 20, 21, 25, 29, 31], available at http://prof.ict.ac.cn/BigDataBench. BDGS supports all the 19 workloads and all six application scenarios in BigDataBench, including Micro Benchmarks, Basic Datastore Operations, Relational Query, Search Engine, Social

Network and E-commerce. We also evaluate BDGS under different experimental settings. Our preliminary experimental results show that *BDGS* can generate big data in a linear gross time as data volume increases while preserving the characteristics of real data.

The rest of the paper is organized as follows. Section 2 presents some basic concepts and requirements in big data generation. Section 3 discusses related work. Section 4 provides an overview of our BDGS. Sections 5 and 6 explain how real-world data sets are selected and synthetic data sets are generated in *BDGS*, respectively. Section 7 reports our preliminary experimental evaluation of BDGS's effectiveness. Finally, Sect. 8 presents a summary and discusses directions for future work.

2   Background and Requirements

Technically, a big data can be broken down into four dimensions: volume, variety, velocity, and veracity [19], which form the 4 V properties of big data.

1. Volume is the most obvious feature of big data. With the exploding of data volume, we now use zettabytes (ZB) to measure data size, developed from petabytes (PB) measurement used just a short time ago.
   IDC forecasts a 44X increase in data volumes between 2009–2020, from 0.8 ZB to 35ZB. From recent Facebook's big data statistics, there are 500+ TB of new data generated per day [12].
2. Velocity is another important but overlooked feature in big data, which denotes data generation, updating, or processing speed. The requirement of generating and processing data quickly gives rise to a great challenge to computer systems performance. According to Twitter's statistics, 340M tweets are sent out everyday, which means 3935 tweets are sent every second.
3. Variety refers to a variety of data types and sources. Comparing to traditional structured data in relational databases, most of today's data are unstructured or semi-structured and they are from different sources, such as movies' reviews, tweets, and photos.
4. Veracity refers to the quality or credibility of the data. With the massive amount of data generated every day, e.g. there are 845M active users on Facebook and 140M active users on Twitter. How to derive trustworthy information from these users and discard noises, and how to generate realistic synthetic data on the basis of raw data, are still open questions.

These properties of big data not only characterize features, but also bring requirements for a new generation of benchmarks. Based on the 4 V properties of big data, we now define the requirements of data generation for benchmarking big data systems. Briefly, big data generators should scale up or down a synthetic data set (volume) of different types (variety) under a controllable generation rate (velocity) while keeping important characteristics of raw data (veracity).

1. Volume: To benchmark a big data system, big data generators should be able to generate data whose volume ranges from GB to PB, and can also scale up and down data volumes to meet different testing requirements.
2. Velocity: Big data applications can be divided into in three types: offline analytic, online service and realtime analytic. Generating input data of workloads to test different types of applications is the basic requirement for data generators. For applications of online services such as video streaming processing, data velocity also means the data processing speed. By contrast, for applications of offline analytic (e.g. k-means clustering or collaborative filtering) and realtime analytic (e.g. select or aggregate query in relational databases), data velocity denotes the data updating frequency. Big data generators should be able to control all the data velocities analyzed above.
3. Variety: Since big data come from various workloads and systems, big data generators should support a diversity data types (structured, semi-structured and unstructured) and sources (table, text, graph, etc.).
4. Veracity: In synthetic data generation, the important characteristics of raw data must be preserved. The diversity of data types and sources as well as large data volumes bring a huge challenge in establishing trust in data generation. A possible solution is to apply state-of-the-art feature selection models, where each model is specialized for one type of data, to abstract important characteristics from raw data. The constructed models can then be used to generate synthetic data. We also note that in model construction or data generation, evaluation are needed to measure the conformity of the model or synthetic data to the raw data.

3   Related Work

How to obtain big data is an essential issue for big data benchmarking. Margo Seltzer, et al. [27] pointed that if we want to produce performance results that are meaningful in the context of real applications, we need to use application-specific benchmarks. Application-specific benchmarking would require application-specific data generators which synthetically scaling up and down a synthetic data set and keeping this data set similar to real data [28]. For big data benchmarking, we also need to generate data of different types and sources. We now review the data generation techniques in current big data benchmarks.

HiBench [18] is a benchmark suite for Hadoop MapReduce. This benchmark contains four categories of workloads. The inputs of these workloads are either data sets of fixed size or scalable and synthetic data sets.

BigBench [16] is a recent effort towards designing big data benchmarks. The data generators in BigBench are developed based on PDGF [26], which is a powerful data generator for structured data. In BigBench, PDGF is extended by adding a web log generator and a review generator. But, in both generators, the veracity of logs and reviews rely on the table data generated by PDGF. In addition, the data generators in BigBench only support the workloads designed to test applications running in DBMSs and MapReduce systems.

Internet and industrial service providers also develop data generators to support their own benchmarks. In LinkBench [9], the social graph data from Facebook are stored in MySQL databases. In this benchmark, the data generator is developed to generate synthetic data with similar characteristics to real social graph data. Specifically, the graph data are first broken into object and association types in order to describe a wide range of data types in the Facebook social graph.

The Transaction Processing Performance Council (TPC) proposes a series of benchmarks to test the performance of DBMSs in decision support systems. The TPC-DS [2] is TPC's latest decision support benchmark that implements a multi-dimensional data generator (MUDD). In MUDD, most of data are generated using traditional synthetic distributions such as a Gaussian distribution. However, for a small portion of crucial data sets, MUDD replies on real data sets to produce more realistic distributions in data generation. Although it can handle some aspects of big data such as volume and velocity, it is only designed for the structured data type.

We summarize the data generation techniques in current benchmarks in Table 1. Overall, existing benchmarks do not adequately address issues relating to keeping the 4 V properties of big data. They are either designed for some specific application domains (e.g. the Hadoop system in Hibench, NoSQL system in YCSB, DBMSs in LinkBench and TPC-DS), or they only consider limited data types (e.g. the structured data type in YCSB and TPC-DS), which do not present the variety of big data. In addition, fixed-size data sets are used as workload inputs in some benchmarks, we call the data volume in these benchmarks "partially scalable". Existing big data benchmarks rarely consider the control of data generation rates (velocity). Hence we call their data generators "uncontrollable" or "semi-controllable" in terms of data velocity. Our BDGS provides an enhanced control of data velocity by developing a mechanism to adjust data

Table 1. Comparison of Big Data Benchmark's data generators

Benchmark's Data Generator	Generator's Extensibility	Volume	Velocity Data Generation Rate	Variety Data Source	Variety Data Type	Variety Software stacks	Veracity
Hibench		Scalable		Text	Unstructured Structured	Hadoop and Hive	Unconsidered
LinkBench		Scalable	Uncontrollable	Table	Structured	Graph database	Partially Considered
CloudSuite	UnExtensible	Partially Scalable		Text, Graph, Table	Structured Unstructured	NoSQL system Hadoop, GraphLab	Partially Considered
TPC-DS		Scalable	Semi-controllable	Table	Structured	DBMS	Unconsidered
YCSB		Scalable	Controllable	Table	Structured	NoSQL system	Partially Considered
BigBench		Scalable	Semi-controllable	Text, Table	Structured Semi-structured Unstructured	Hadoop and DBMS	Partially Considered
BigDataBench	Extensible	Scalable	Semi-controllable	Text, Graph, Table	Structured Semi-structured Unstructured	NoSQL system DBMS Realtime Analytics Offline Analytics systems	Considered

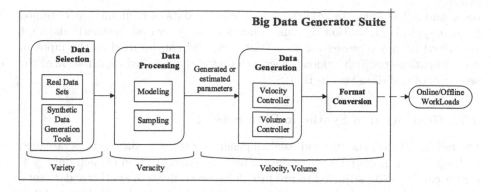

Fig. 1. The Architecture of BDGS

generation speeds. Meanwhile, most of benchmarks employ random data sets as workload inputs, which do not consider the veracity of big data. In addition, some data generation tools use similar distributions with real scenarios (data veracity is partially considered). Hence using such synthetic data, it is difficult to make accurate testing of realistic application scenarios; that is, current synthetic workload inputs incur a loss of data veracity.

4   The Methodology of BDGS

In this section, we present an overview of BDGS that is implemented as an component of our big data benchmark suite BigDataBench [29]. Briefly, BDGS is designed to provide input data for application-specific workloads in BigData-Bench. As shown in Fig. 1, the data generation process in BDGS consists of four steps.

The first step is data selection, which reflects the variety of big data, in which the main operation is the selection of representative real-world data sets or tools to generate synthetic data sets. The second step is data processing, the chosen data sets are processed to exact their important characteristics, thus preserving the veracity of data. Sampling methods are also provided to scale down data sets. The third step is data generation, the volume and velocity are specified according to user requirements before a data set is generated. Finally, the format conversion tools transform the generated data set into a format capable of being used as the input data of a specific workload. At present, we are working on implementing a parallel version of BDGS.

4.1   Selection of Data Sets

In BDGS, we select the real data sets with the following considerations. First, real data sets should cover different data types including structured, semi-structured and unstructured data, as well as different data sources such as text data, graph data and table data. In addition, each chosen data set should be obtained from a typical application domain. Therefore, a generated synthetic data set is derived from a representative real data set, and can be used as the input of an application-specific workload. Section 5 provides a detailed explanation of the selection of real data sets in BDGS.

4.2   Generation of Synthetic Data Sets

Overall, in BDGS, the method used to generate synthetic data sets is designed to keep the 4 V properties of big data. First, the configuration of each data generator can be easily adjusted to produce different volumes of synthetic data sets. Similarly, different velocities of data sets can be generated by deploying different numbers of parallel data generators. In BDGS, different kinds of generators are developed to generate a variety of data sets with different types and sources. Finally, the synthetic data sets are generated using the data models derived from real data sets, thus preserving the data veracity in generation. Section 6 provides a detailed explanation of the synthetic data generation in BDGS.

5   Real-World Data Sets

Our data generation tools are based on small-scale real data, this section introduces the representative real data sets we selected. To cover the variety and veracity, we choose six representative real data sets, which can be download from http://prof.ict.ac.cn/BigDataBench/#downloads. As shown in Table 2, our chosen real data sets are diverse in three dimensions: data type, data source, and application domain. In the these data sets, the Wikipedia Entries, Google Web Graph, and Facebook Social Graph are unstructured; E-commence transaction data are structured; and Amazon Movie Reviews and Personal Resumes are semi-structured. Specifically, Wikipedia Entries are plaint text, and Amazon Movie Reviews are text but have some schema constrains. Google Web Graph are directed graphs in which nodes are web pages and each edge denotes a linking of two pages. Facebook Social Graph are undirected graphs in which two persons are friends of each other. As we shown later, Amazon Movie Reviews can be converted to bipartite graphs in which each edge has attributes like score and review text. E-commence transaction data are typical relational tables, and Personal Resumes can also be seen as table-like data with less schema. Finally, our chosen data sets are from different internet service domains: Wikipedia Entries and Google Web Graph can be used for the search engine domain, while Personal Resumes are from the vertical search engine. Amazon Movie Reviews and E-commence transaction are from the E-commence domain, and Facebook Social Graph are from the social network domain.

Table 2. The summary of six real life data sets.

Data sets	Data type	Data source	Data size
Wikipedia Entries	un-structured	text data	4,300,000 English articles
Amazon Movie Reviews	semi-structure	text data	7,911,684 reviews
Google Web Graph	un-structured	graph data	875713 nodes, 5105039 edges
Facebook Social Network	un-structured	graph data	4039 nodes, 88234 edges
E-commence Transaction	structured	table data	Table 1: 4 columns, 38658 rows. Table 2: 6 columns, 242735 rows
Person Resumes Data	semi-structured	table data	278956 resumes

1. **Wikipedia Entries** [8]. The Wikipedia data set is unstructured, with 4,300,000 English articles.
2. **Amazon Movie Reviews** [3]. This data set is semi-structured, consisting of 7,911,684 reviews on 889,176 movies by 253,059 users. The data span from Aug 1997 to Oct 2012. The raw format is text, and consists of productID, userID, profileName, helpfulness, score, time, summary and text.
3. **Google Web Graph** [5]. This data set is unstructured, containing 875713 nodes representing web pages and 5105039 edges representing the links between web pages. This data set is released by Google as a part of Google Programming Contest.
4. **Facebook Social Graph** [4]. This data set contains 4039 nodes, which represent users, and 88234 edges, which represent friendship between users.
5. **E-commence Transaction**. This data set is from an E-commence web site, which is structured, consisting of two tables: ORDER and order ITEM.
6. **Personal Resumes**. This data is from a vertical search engine for scientists developed by ourselves. The data set is semi-structured, consisting of 278956 resumes automatically extracted from 20,000,000 web pages of university and research institutions. The resume data have fields of name, email, telephone, address, date of birth, home place, institute, title, research interest, education experience, work experience, and publications. Because the data are automatically collected from the web by our program, they are not normalized: some information may be omitted in the web pages, while others may be redundant. For example, a person may only list name, email, institute, title while keeping other information blank, and he may have two emails.

6 Synthetic Data Generators

This section presents our big data generator suite: BDGS, which is an implementation of our generic data generation approach in Fig. 1. This implementation includes six data generators belonging to three types: Text Generator, Graph Generator and Table Generator. BDGS can generate synthetic data while preserving the important characteristics of real data sets, can also rapidly scale data to meet the input requirements of all the 19 workloads in BigDataBench.

6.1   Text Generator

Based on the chosen text data sets such as Wikipedia entries, we implement our text generator. It applies latent dirichlet allocation (LDA) [11] as the text data generation model. This model can keep the veracity of topic distributions in documents as well as the word distribution under each topic.

In machine learning and natural language processing, a topic model is a type of statistics designed to discover the abstract topics occurring in a collection of documents [7]. LDA models each document as a mixture of latent topics and a topic model is characterized by a distribution of words. The document generation process in LDA has three steps:

1. choose $N \sim \text{Poisson}(\xi)$ as the length of documents.
2. choose $\theta \sim \text{Dirichlet}(\alpha)$ as the topic proportions of document.
3. for each of N words w_n:
   (a) choose a topic $z_n \sim \text{Multinomial}(\theta)$
   (b) choose a word w_n from $p(w_n|, z_n, \beta)$, a multinomial probability conditioned on the topic z_n

Figure 2 shows the process of generating text data. It first preprocesses a real data set to obtain a word dictionary. It then trains the parameters of a LDA model from this data set. The parameters α and β are estimated using a variational EM algorithm, and the implementation of this algorithm is in lda-c [6]. Finally, we use the LDA process mentioned above to generate documents.

Since LDA models both word and hidden topics in text data, more characteristics of real raw data can be preserved than just applying the word-level modeling. Documents are independent of each other, so the generation of different documents can be paralleled and distributed using multiple generators under the same parameters, thus guaranteeing the velocity of generating data. Users can also configure the number of documents or the size of text data, thus generating different volumes of text data.

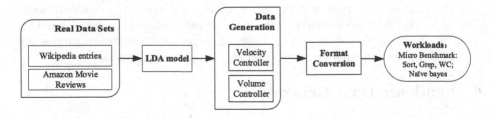

Fig. 2. The process of Text Generator

6.2   Graph Generator

A graph consists of nodes and edges that connect nodes. For the graph data sets, namely Facebook Social Graph and Google Web Gragh, we use the kronecker graph model in our graph generator. Using a small number of parameters, the kronecker graph model [23] can capture many graph patterns, e.g. the Denazification Power Law and the Shrinking Diameters effect.

Some temporal properties of real networks, such as densification and shrinking diameter can also be cover. We believe that the kronecker graph can effectively model the structure of real networks. Many temporal properties of real networks such as densification and shrinking diameter can also be covered by the kronecker graph [22], thus guaranteeing the veracity of the generated graph data.

The kronecker graph model is designed to create self-similar graphs. It begins with an initial graph, represented by adjacency matrix N. It then progressively produces larger graphs by kronecher multiplication. Specifically, we use the algorithms in [22] to estimate initial N as the parameter for the raw real graph data and use the library in Stanford Network Analysis Platform (SNAP, http://snap.stanford.edu/) to implement our generator. Users can configure the number of nodes in the graphs, so the graph volume can be scaled up. The kronecker graph can be generated in linear time with the expected number of edges. Figure 3 shows the process of our Graph Generator.

Although the raw format is text, the Movie Reviews data set can also been seen as a graph. As shown in Fig. 4, the left nodes represent users (each user has a Id), and the right nodes represent products (each product has a Id). The edge means the left user commented review on the right product, while the attributes of the edge are score and review text.

Movie Reviews data set can be used for two workloads: collaborative filtering and sentiment classification. For collaborative filtering, we use the productId, userId and score which can be seen as user-product score matrices. The task of collaborative filtering is to predict the missing score in the matrices. While sentiment classification uses the review text as the input, and the score as the category label. As a result, when we generate the big synthetic review data, we only concern the fields of productId, userId, score and text.

There are two steps to generate Movie Reviews data. First, applying graph generator to generate a bipartite graph. Second, for each edge in the graph, using

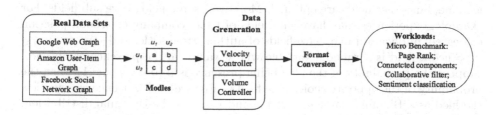

Fig. 3. The process of Graph Generator

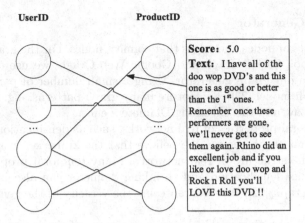

Fig. 4. The process of generating Review data

the multinational random number generator to generate a score S, then using the trained LDA under category S to generate review text for this edge.

6.3   Table Generator

To scale up the E-commence transection table data, we use the PDGF [26], which is also used in BigBench and TPC-DS. PDGF uses XML configuration files for data description and distribution, thereby simplifying the generation of different distributions of specified data sets. Hence we can easily use PDGF to generate synthetic table data by configuring the data schema and data distribution to adapt to real data.

Personal Resumes data are also a kind of table-like data (semi-structured data). We implement a resume generator to generate personal resumes for basic data store operations that are similar to those of YCSB. Traditionally, the input data in YCSB are table records that set all the fields as mandatory, however, the real world data especially in NoSQL scenarios are schema less. That is to say, the records can have arbitrary fields. To model these uncertainties of real data, we use the professor resumes in ProfSearch (available at http://prof. ncic.ac.cn/) as the real data set. A resume may consist of name, email, telephone, address, date of birth, home place, institute, title, research interest, education experience, work experience and publications. Each resume's primary key is name, but other fields are optional. Moreover, some fields have sub fields. For example, experiences may have sub fields of time, company or school, position or degree, publication may have fields of author, time, title, and source.

Since a field is a random variable whose value can be presence or absence, suppose the probability of the filed being presence is p $(0 <= p <= 1)$, then the probability of the opposite choice (the filed is absence) is 1-p. Hence we can view the filed as a Binomial random variable and it meets the Binomial distribution. As a result, we generate personal resume data using the following three steps. Fig. 5 shows the process of resume generator.

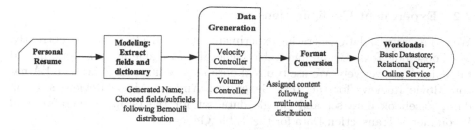

Fig. 5. The process of generating resume data.

1. Randomly generate a string as the name of a resume
2. Randomly choose fields from email, telephone, address, date of birth, home place, institute, title, research interest, education experience, work experience and publications, where each field follows the bernoulli probability distribution.
3. For each field:
   if the field has sub fields, then randomly choose its sub fields following the bernoulli probability distribution. **else** assign the content of the field using the multinomial probability distribution.

7 Experimental Evaluation

This section first describes the metric to evaluate our BDGS and the experimental configurations, following the results of experimental evaluation. The evaluation is designed to illustrate the effectiveness of our BDGS in generating different types of synthetic data sets based on real data sets from different sources. The evaluation results also shows the consumed generation time with respect to the volume of generated data.

We implemented BDGS in Linux environment, in which most of programs are written in C/C++ and shell scripts. Although the current experiments were doing on a single node, the generators and schemas can be deployed in multiple nodes.

7.1 Metric

To evaluate the effectiveness of our methodology and our big data generator suite, we use a user-perceived performance metric–data generation rate to evaluate the data velocity under different data volumes. The data generation rate is defined as the amount of data generated divided by the running time for generation data. We use Edgs/s as the generation rate for graph data and MB/s as the generation rate for all the other data types. For examples, in graph data generation, generating 100000 edges in 100 seconds means the generation rate is 100000/100=1000 Edgs/s. In text data generation, generating 100 GB (i.e. 102400 MB) text data in 10000 seconds means the generation rate is 102400/10000=10.24MB/s.

7.2   Experiment Configurations

We ran a series of data generating experiments using BDGS, in which, currently, we choose our Text, Graph and Table Generators to generate text, graph, and table data, respectively. We use two real data sets: Wikipedia Entries and Amazon Movie Reviews for the Text Generator; Amazon Movie Reviews's graph data, Facebook data set and Google data set for the Graph Generator; and E-commerce Transaction Data for the Table Generator.

In experiments, we generated big data on one node with two Xeon E5645 processors equipped with 32GB memory and 8X1TB disk. In generation, the data size ranges from 10 GB to 500 GB for the Text Generator and the Table Generator, and 2^{16} to 2^{20} for the Graph Generator.

7.3   Evaluation of BDGS

We tested our Text, Graph and Table Generators under different data volumes and employed the data generation rate to evaluate these generators. Figures 6-1, 7-1 and 8-1 report the data generation rates of the Text, Graph and Table Generator, respectively. In the Text Generator, the average generation rate of the Wiki data set is 63.23 MB/s, and this rate of Amazon movies reviews(5 scores) data set is 71.3 MB/s. The difference is caused by the dictionary's size, which is 7762 in Wiki data set, and 5390 in Amazon data set. The least average generation rate of the Graph Generator under enough memory is 591684 Edges/s, and this rate of the Table Generator is 23.85 MB/s.

Furthermore, the experiment results in Figs. 6-2, 7-2 and 8-2 show that our generators can rapidly generate big data in a linear gross time with the data volume increases. Hence, it is reasonable to say that our generators can maintain a roughly constant rate to generate high volume data for different data types. For example, generating 1 TB Wiki data takes 4.7 h.

In addition, we can observe in Figs. 6, 7 and 8 that although the data generation rates roughly maintain constants, these rates have slightly changes when dealing with different data types. For the Text Generator and the Graph Generator, the data generation rate varies slightly in spite of which real data sets

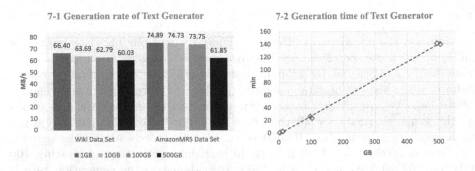

Fig. 6. Data Generation Rates and Time of the Text Generator

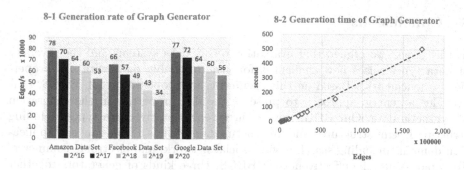

Fig. 7. Data Generation Rates and Time of the Graph Generator

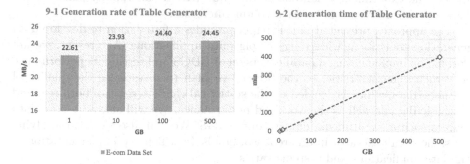

Fig. 8. Data Generation Rates and Time of the Table Generator

are used, and this variation is mainly caused by the high volume of data and the limitation of memory capacity. Based on the observation that the memory resource utilization is above 90 % in experiments, we believe the performance bottleneck of the Text Generator is memory. The next edition should pay more attention to optimize memory control. In addition, since the Graph Generator needs to compute the whole Graph in memory to match the real data sets and generate a big map, the Graph Generator has a larger memory requirement. This incurs the smallest data generation rate in the Graph Generator. Finally, in the Table Generator the data generation rate slightly increases with the data volume; that is, the average execution time to generate a unit of data decreases. This is because the total execution time of the Table Generator consists of a long configuration time and the data generation time. Given that the generation time per unit of data keeps fixed when the data volume increases, the average configuration time per unit of data (i.e. the total configuration time divided by the total data size) deceases as data volume increases. Therefore, the execution time per unit of data (i.e. the sum of the data generation time per unit of data and the average configuration time per unit of data) decreases.

To summarize, our big data generator suite can rapidly generate big data in a linear gross time as data volume increases. It not only covers various data types and sources, but also considers generating big data on the basis of representative real data sets, thereby preserving the characteristics of real data.

8 Conclusions

In this paper, we argued that evaluation of big data systems and the diversity of data types raises new challenges for generating big data with 4 V properties in benchmarks. Based on this argument, we proposed a tool, BDGS, that provides a generic approach to preserve the 4 V properties in the generation of synthetic data. Our BDGS covers three representative data types including structured, semi-structured and unstructured, as well as a wide range of application domains including search engine, social network, and electronic commerce. To demonstrate the effectiveness of BDGS, three kinds of generators together with their data format conversion tools were developed and experimentally evaluated. The experiment results show that our BDGS can rapidly generate big data in a linear gross time as data volume increases.

The approach presented in this paper provides a foundation to develop data generation tools for a wide range of big data applications. At present, we are working on implementing a parallel version of BDGS and testing its performance on veracity characteristic. In the future, we plan to investigate the quality of synthetic data, which is decided by the statistical characteristic of both real and synthetic data as well as the workload performance. We will also add more data sources such as multimedia data to our BDGS. We will also work on applying the same data generator in different workloads, in which the implementation of big data applications and systems varies.

References

1. http://www-01.ibm.com/software/data/bigdata/
2. http://www.tpc.org/tpcds/
3. Amazon movie reviews. http://snap.stanford.edu/data/web-Amazon.html
4. Facebook graph. http://snap.stanford.edu/data/egonets-Facebook.html
5. Google web graph. http://snap.stanford.edu/data/web-Google.html
6. Lda-c home page. http://www.cs.princeton.edu/blei/lda-c/index.html
7. Topic model. http://en.wikipedia.org/wiki/Topic_model
8. wikipedia. http://en.wikipedia.org
9. Armstrong, T.G., Ponnekanti, V., Borthakur, D., Callaghan, M.: Linkbench: a database benchmark based on the facebook social graph. In: SIGMOD'13 (2013)
10. Barroso, L.A., Hölzle, U.: The datacenter as a computer: an introduction to the design of warehouse-scale machines. Synth. Lect. Comput. Archit. 4(1), 1–108 (2009)
11. Blei, D.M., Ng, A.Y., Jordan, M.I.: Latent dirichlet allocation. J Mach. Learn. Res. 3, 993–1022 (2003)
12. Fourneau, J.-M., Pekergin, N.: Benchmark. In: Calzarossa, M.C., Tucci, S. (eds.) Performance 2002. LNCS, vol. 2459, pp. 179–207. Springer, Heidelberg (2002)
13. Ferdman, M., Adileh, A., Kocberber, O., Volos, S., Alisafaee, M., Jevdjic, D., Kaynak, C., Popescu, A.D., Ailamaki, A., Falsafi, B.: Clearing the clouds: a study of emerging workloads on modern hardware. In: Proceedings of the 17th Conference on Architectural Support for Programming Languages and Operating Systems, ASPLOS 2012, pp. 1–11 (2011)

14. Gao, W., Zhu, Y., Jia, Z., Luo, C., Wang, L., Li, Z., Zhan, J., Qi, Y., He, Y., Gong, S., et al.: Bigdatabench: a big data benchmark suite from web search engines. In: The Third Workshop on Architectures and Systems for Big Data (ASBD 2013), in conjunction with ISCA 2013 (2013)
15. Ghazal, A.: Big data benchmarking-data model proposal. In: First Workshop on Big Data Benchmarking, San Jose, Califorina (2012)
16. Ghazal, A., Rabl, T., Hu, M., Raab, F., Poess, M., Crolotte, A., Jacobsen, H.-A.: Bigbench: towards an industry standard benchmark for big data analytics. In: SIGMOD, ACM (2013)
17. Gray, J., Sundaresan, P., Englert, S., Baclawski, K., Weinberger, P.J.: Quickly generating billion-record synthetic databases. In: ACM SIGMOD Record, vol. 23, pp. 243–252. ACM (1994)
18. Huang, S., Huang, J., Dai, J., Xie, T., Huang, B.: The hibench benchmark suite: Characterization of the mapreduce-based data analysis. In: 2010 IEEE 26th International Conference on Data Engineering Workshops (ICDEW), pp. 41–51. IEEE (2010)
19. IBM. http://www.ibm.com/developerworks/bigdata/karentest/newto.html
20. Jia, Z., Wang, L., Zhan, J., Zhang, L., Luo, C.: Characterizing data analysis workloads in data centers. In: IEEE International Symposium on Workload Characterization (IISWC), IEEE (2013)
21. Jia, Z., Zhou, R., Zhu, C., Wang, L., Gao, W., Shi, Y., Zhan, J., Zhang, L.: The Implications of diverse applications and scalable data sets in benchmarking big data systems. In: Rabl, T., Poess, M., Baru, C., Jacobsen, H.-A. (eds.) WBDB 2012. LNCS, vol. 8163, pp. 44–59. Springer, Heidelberg (2014)
22. Leskovec, J., Chakrabarti, D., Kleinberg, J., Faloutsos, C., Ghahramani, Z.: Kronecker graphs: an approach to modeling networks. J. Mach. Learn. Res. 11, 985–1042 (2010)
23. Leskovec, J., Chakrabarti, D., Kleinberg, J.M., Faloutsos, C.: Realistic, mathematically tractable graph generation and evolution, using kronecker multiplication. In: Jorge, A.M., Torgo, L., Brazdil, P.B., Camacho, R., Gama, J. (eds.) PKDD 2005. LNCS (LNAI), vol. 3721, pp. 133–145. Springer, Heidelberg (2005)
24. Lotfi-Kamran, P., Grot, B., Ferdman, M., Volos, S., Kocberber, O., Picorel, J., Adileh, A., Jevdjic, D., Idgunji, S., Ozer, E., et al.: Scale-out processors. In: Proceedings of the 39th International Symposium on Computer Architecture, pp. 500–511. IEEE (2012)
25. Luo, C., Zhan, J., Jia, Z., Wang, L., Zhang, L., Sun, N.: Cloudrank-d: benchmarking and ranking cloud computing systems for data processing applications. Front. Comput. Sci. 6(4), 347–362 (2012)
26. Rabl, T., Frank, M., Sergieh, H.M., Kosch, H.: A Data generator for cloud-scale benchmarking. In: Nambiar, R., Poess, M. (eds.) TPCTC 2010. LNCS, vol. 6417, pp. 41–56. Springer, Heidelberg (2011)
27. Seltzer, M., Krinsky, D., Smith, K., Zhang, X.: The case for application-specific benchmarking. In: Proceedings of the Seventh Workshop on Hot Topics in Operating Systems, 1999, pp. 102–107. IEEE (1999)
28. Tay, Y.C.: Data generation for application-specific benchmarking. In: VLDB, Challenges and Visions (2011)
29. Wang, L., Zhan, J., Luo, C., Zhu, Y., Yang, Q., He, Y., Gao, W., Jia, Z., Shi, Y., Zhang, S., Zhen, C., Lu, G., Zhan, K., Qiu, B.: Bigdatabench: A big data benchmark suite from internet services. In: The 20th IEEE International Symposium on High-Performance Computer Architecture(HPCA) (2014)

30. Zhan, J., Zhang, L., Sun, N., Wang, L., Jia, Z., Luo, C.: High volume comput-
    ing: Identifying and characterizing throughput oriented workloads in data centers.
    In: 2012 IEEE 26th International Parallel and Distributed Processing Symposium
    Workshops & PhD Forum (IPDPSW), pp. 1712–1721. IEEE (2012)
31. Zhu, Y., Zhan, J., Weng, C., Nambiar, R., Zhang, J., Chen, X., Wang, L.: Gen-
    erating comprehensive big data workloads as a benchmarking framework. In: The
    19th International Conference on Database Systems for Advanced Applications
    (DASFAA 2014) (2014)

A Multidimensional OLAP Engine Implementation in Key-Value Database Systems

Hongwei Zhao[✉] and Xiaojun Ye

School of Software, Tsinghua University, Beijing 100084, China
hwzhao2012@gmail.com, yexj@mail.tsinghua.edu.cn

Abstract. This paper tries to explore the capabilities of MapReduce-like execution engines for multidimensional data analytics through implementing a Multidimensional Online Analytical Processing (MOLAP) engine with cube model on the Hadoop ecosystem. The cube storage module converts dimension members into binary keys and leverages a novel distributed database to provide efficient storage for huge cuboids. The bit encoding sparse index is used to compress the cube data and the dimension bit encoding key with maximum members is used to achieve cube data sharding and distributed aggregation computing. We discuss how to match the star schema with cube model databases, and how to use resilient distributed data-sets for MDX-like queries executing on key-value systems. Finally, some queries of TPC-DS benchmark are adopted to validate the prototype implementation of the MOLAP engine. The results indicate that designed scenarios based on TPC-DS are suitable for various big data analytics operation benchmarking.

Keywords: Cloud computing · Multidimensional on-line analysis processing · Cube model · TPC-DS benchmark

1 Introduction

Online Analytical Processing (OLAP), the core component of Business Intelligence (BI) systems, is often built on a Relational Database Management System (RDBMS) to further tune and optimize analytical queries. Such OLAP systems face a confluence of growing challenges deriving from the latest big data revolution [1]. Scalability and flexibility challenge earlier-generation OLAP technologies and architectures [2].

Although big data analytics techniques over large-scale data repositories have been investigated recently on Hadoop ecosystem technologies, such as Dremel [3], Spanner [4], Shark and Spark [5], extending data warehouse components like report, dashboard, and specifically OLAP engine on these heterogeneous cloud platforms plays a leading role for future business analytics.

The purpose of OLAP engines is to provide OLAP algebra operators such as roll-up/drill-down, slice/dice or pivot capabilities on data warehouse systems. But traditional OLAP engines have poor scalability and they cannot handle big scale of data easily. They have to take an evolutionary architecture employing distributed and

T. Rabl et al. (Eds.): WBDB 2013, LNCS 8585, pp. 155–170, 2014.
DOI: 10.1007/978-3-319-10596-3_12

parallel computing technologies. Compared to ROLAP (Relational OLAP) data models which need many joins, MOLAP (multidimensional OLAP) are more suitable for cloud computing where cube data (cuboids) are naturally stored in key-value databases. As such, MOLAP based on NoSQL systems can easily be partitioned, distributed, and scaled out for big data analytic and fill the gap between big data stores and rapid interactive business analytic demands [1, 2].

In this paper, we present a MOLAP engine prototype, a key-value database performance testing framework implementation for persistent cube data according to the business scenarios of TPC-DS benchmark. We aim at integrating well-known benefits of cube model into big data infrastructure in order to achieve more powerful analytics capabilities. Inspired by resilient distributed dataset (RDD), we create RDDs for cube blocks, and implement OLAP operators to aggregate and query cube data based on in-memory RDD to achieve better performance.

The main contributions of this paper are to answer how to build cube data and store them into key-value databases and how to extend typical OLAP operators in MapReduce-like executions on cloud systems. Key points are summarized as follows.

1. The OLAP engine is built on the Hadoop ecosystem to embrace large scale data repositories. We choose HBase for distributing cube data storage and improve stability by its fine-grained fault tolerance model.
2. The cube data are constructed into a multidimensional array, which is linearized with BESS [6] method that is suitable for cube data shard in the key-value storage. The query plan can be split into smaller aggregation computing tasks for concurrent running on distributed work nodes.
3. The system is implemented using Akka actor model [7], a framework to allow building an event-driven concurrent system. We use a dispatcher actor and a bunch of managed remote or local worker actors to fulfill the scalability demands. This engine framework also provides different levels of concurrency.

The rest of the paper is structured as follows: Sect. 2 briefly reviews past data analytics technology research in business intelligence on big data and related OLAP benchmarking. A detailed description of the proposed OLAP engine architecture is presented in Sect. 3. According to the demand driven cube design methodology, the snowflake schema of TPC-DS is converted into a multidimensional model. Based on linearization of multidimensional array, we present the cube logical structure and schema for key-value storage. In Sect. 4, testing scenarios are used to identify various performance criteria based on TPC-DS benchmark dataset. Finally, In Sect. 5 we conclude the paper.

2 Related Work

2.1 Business Intelligence on Big Data

Nowadays main database vendors have provided their own NoSQL products (Windows SQL Azure, Oracle NoSQL) or Hadoop data integration and business analytics platforms (Oracle Big Data Appliance, IBM InfoSphere) that can be integrated with their existing data warehouses for big data analytics solutions.

In typical SQL scenarios for big data analysis on Hadoop ecosystem technologies there are Hive, Cloudera Impala or Google Dremel as ROLAP tools [8]. ROLAP is designed to leave the data where it is and defer to process until it is actually queried. It becomes slow because running complex MapReduce jobs that consist of grouping and joining large data sets. The pure MapReduce framework executes SQL queries on partially aggregated data which can be too slow for big data browsing. MapReduce-like ROLAP engines lack features that make data analytics interactive.

Therefore, we choose MOLAP rather than ROLAP as the underlying physical model for big data analytics benchmarking. MOLAP stores data in multidimensional arrays. Each dimension of the array represents the respective dimension of the cube. The contents of the array are the measure(s) of the cube. Conceptually, the data cube consists of huge cuboids, the finest granularity view is base cuboids which contain the full complement of dimensions (or attributes), surrounded by a collection of 2^{d-1} cuboids that represent the aggregation of the base cuboid along one or more dimensions.

The problem of dimensional explosion causes the whole cube data size to increase dramatically. To minimize the storage requirements of cubes in key-vale stores, we can create the base cuboids in the cube building phase (testing data population). When aggregation cuboids are needed by user queries, they are produced step-by-step and cached during data analytics process (query stage). As such, only base cuboid data are stored in distributed file systems to keep space usage efficient, aggregation cuboids are created in memory by user query requests. This method results in the performance balance with cube data storage scale.

Figure 1 presents the data store structure of our designed MOLAP where metadata are derived from cube lattice and cube instance data are divided into dimension instance data and cuboid instance data. According to the cube lattice, the base cuboids

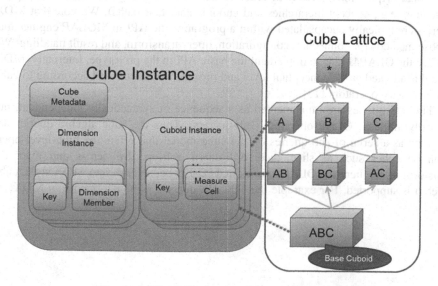

Fig. 1. MOLAP database on hadoop ecosystems

are used to calculate all of the aggregation cuboids. In the cube building stage the base cuboid can be populated from extern relational databases. In query stage, other cuboids queried will be computed and stored for next uses by MOLAP operations. Resilient Distributed Datasets (RDDs) [16] can be created in each node that holds a part of cuboids in memory. Then they can be used to produce the high-level cuboids with user distributed aggregation queries.

2.2   Cube Algebra with TPC-DS Queries

In the following, we use TPC-DS dataset to build MOLAP cubes. TPC-DS model contains the decision support functions of a retailer with 7 fact tables and 17 shared dimensions. Its 99 queries are classified into four classes: reporting queries, ad hoc decision support queries, interactive OLAP queries, and extraction queries [9].

We can consider cube modeling from two baselines provided by TPC-DS: (1) According to Kimball's "first principles" design method, the design of cubes should be based on analysis of user query requirements, e.g. based on the knowledge of the application area and the types of queries the users are expected to pose [10]. (2) Like cube construction based on queries proposed in [11], the cube construction may be also based on the use of functional dependency information in TPC-DS data schema.

In this paper, we use TPC-DS testing query templates as a logical cube design method [12]. To construct a cube, we detect measures and dimension members used in TPC-DS queries' templates. Our target is to build a maximized cube in order to answer all related queries. It should be also minimized to assure not to contain other dimensions and measures except for the needed ones.

It is difficult for SQL queries to effectively exploit cube specific constructs. Thus, we work with existing TPC-DS queries to transform them from SQL to MultiDimensional eXpressions (MDX) in order to take full advantage of OLAP conceptual structures (e.g., concept hierarchies and cuboids and their paths). We note that MDX queries are typically encapsulated within a programmatic API in MOLAP engines that expose methods for connection configuration, query transform, and result handling. We can use the OLAP4J [24] to implement the basic API in the prototype. Internally, MDX model is adopted at the conceptual level and provide a common API according to cube algebra query operators [13].

TPC-DS queries are implemented as a sequence of syntactically independent, but logically affiliated OLAP operators (sub-queries). Thus the cube algebra queries are regarded as a serial of sub-queries on the cube model that offers the high-level operation on cubes such as slice, dice, drill-down, roll-up, drill-across and map. By transforming the iterative OLAP queries in TPC-DS benchmark into MDX style, cube algebra is supported. For example the TPC-DS query 7 is transformed into MDX:

```
select { i_item_id } on rows,
  { avg(ss_quantity), avg(ss_list_price),
    avg(ss_coupon_amt), avg(ss_sales_price) }   on columns
from store_sales_cube
where (cd_gender .[Male],
       cd_marital_status .[Single],
       cd_education_status .[College],
       d_year.[2000])
```

Based on the cube model, a set of OLAP algebras core common operations for TPC-DS can be identified and implemented in MOLAP engines on key-value systems. These cube operations are selection, projection, and drill across, union/intersection/difference, change level, change base, pivot, etc.

3 System Design

As stated before, MOLAP data storage will be needed to address the scalability issue of data management. We use a key-value database to store the compressed array cube. The logic model is based on the concept of the base cuboids representing the most detailed information (i.e. the information at the lowest levels of the cube lattice). All other cuboids are calculated from the base cuboids. All the cuboids are mapped to multidimensional arrays with the dimension coordinators encoded into bit keys. The loading method for the hash-based base cuboids is better for larger data sets under distributed environments [14]. Thus we use a designed hash key algorithm and implement the engine with Akka actor model that run on the HBase.

The MOLAP engine architecture, as shown in Fig. 2, can run in a large cluster in which there are two roles: one is Dispatcher Node, which maintains MDX query task status and dispatches OLAP operation tasks; the other is Worker Node that executes diverse OLAP operation tasks and caches the partition cuboids in-memory. The engine, built on the concept of in-memory cluster [15], provides several array-based primitives to manipulate the entire set of fragments associated with the data cube. Some relevant examples include: (i) data aggregation (i.e. max, min, average), (ii) data sub-setting (slicing and dicing), (iii) data cube rolling-up and drilling-down, (iv) data cube pivoting etc.

Key implementation features differentiate the engine from traditional OLAP engines on relational databases:

1. With Akka actor topology, it can run MDX queries in a MapReduce-like style that supports several stages to extract and transform the cube data.
2. With an array structure and key-value compression, it reduces both the data size and the processing time while naively storing the data in its original format as far as possible.
3. It is optimized for low latency, and an in-memory distributed storage for cube data is provided which lets the application keep it in memory across queries [15].

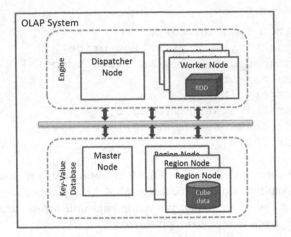

Fig. 2. System architecture

3.1 Demand Driven Cube Modeling

We use TPC-DS's queries to illustrate the demand-driven cube model construction approach [16]. This method produces star schemas from the requirements (i.e. MDX queries). Moreover, this method is able to cope with denormalization in the input relational schemas and get equivalent outputs, which means it can change denormalization relation data into the multidimensional data.

Cube models are generated in two phases: cube building and cube merging, so that it is derived along two stages for TPC-DS performance testing:

1. For each input query, the first stage extracts the multidimensional knowledge contained in the query according to the predefined graph of data table relationship. Along this stage, the role played by the graph will be crucial to infer the base cuboid's metadata.
2. The second stage merges each multidimensional graph according to multidimensionality. To do so, this method defines a set of cuboids that must be preserved in order to summarize themselves into a base cuboid. This step is to find the completed and minimized cube model.

In the following, we focus on interesting TPC-DS queries and evaluate the associated cube data only. We materialize all of the base cuboids that are inferred from testing queries. Cuboids generated are decided by the related queries.

3.2 Cube Logical Structure Generation

A multidimensional array to organize star schema data is provided in [17]: measures are stored in a separate array as a cuboid data, and each dimension is used to form dimensional arrays. We convert the linearized array into bit key array by bit-encoded sparse structure (BESS) [6] (see Fig. 3). Each member in dimension d_i can be

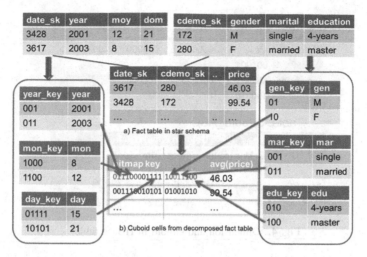

Fig. 3. Bit key encoding and combination from TPC-DS fact table

represented by an integer value between 0 and n_i. The integer is converted into bit key in the length of $\log n_i$. BESS concatenates each bit key of dimension member as the index of cuboids. The performance is optimized to use logic operations to replace multiplication and division operations.

For example, there are three attributes in the table date_dim: year, month and day in TPC-DS data schema. Meanwhile, there are 7 values in year, 12 values in month and 31 values in day. Therefore, year is encoded with 3 bits, e.g. 2001 as 001, month with 4 bits, e.g. 12 as 1100, day with 5 bits. The length of date_dim bit key is 12 bits and each dimension has its own mask: dimension year's mask is 111000000000, month is 000111100000, day is 000000011111.

In this linearized function, each dimension has a mask to allocate its effective bits in the key. When new members are added, the mask should be expanded if the dimension member size has reached the maximum of the effective bits.

3.3   Cube Physical Storage Implementation

The cube data are composed of two parts: dimension data and cuboid data. We choose HBase, a NoSQL database [18], to store cube data into two tables separately.

The dimensions and dimension hierarchies are explicitly stored with dimension name as row key. Furthermore, the dimension member and its bit key are stored as key-value pairs. The reversed pairs are also stored as indices and used for the query executing that needs to convert the where conditions into filter bit key.

The cube's measure data composed of different dimensions are stored in each cuboid. The bit key for the cell of cuboid is taken as the row key of the HBase table. The row key in HBase is sorted and partitioned into data file(s) in a region server. And the field and aggregation name are combined into one string as the qualifier, for

Table: Dimension

Row Key	Column Family: default			
Dimension A	Mask	000001	001000	001001
	001001	A_1	A_2	A_3
Dimension B	Mask	000010	100000	
	100010	B_1	B_2	

Table: Cuboid_ABC

Row Key	Column Family: default		
000111	Mea_count	Mea_sum	
	1	M_1	
011010	Mea_count	Mea_sum	
	1	M_2	

Fig. 4. Dimension and cuboid in HBase table

example, M1_count, M1_sum. The measures can be added or updated by the HBase native Put command. The aggregation value for this cuboid is stored as the value of an HBase cell (Fig. 4).

3.4 Cube Data Shard and Distributed Aggregation Computation

When considering cube data shards, the horizontal round-robin partitioning approach is the simplest strategy. After fetching the joined fact data, entire records are uniformly distributed onto worker nodes, according to the involved dimension with the highest cardinality. For example, the cube built from TPC-DS Query 7 has the largest dimension item_id. With n worker nodes in the cluster, the dispatcher will send the ith cuboid to the node (i mod n). This strategy enables the sequential access to a node to be done in parallel.

By this way, each node aggregates the measures during the cube building. Since there are no points of resource contention between nodes, this allows aggregation functions on the nodes to be carried out in a faster manner. We can perform distributed aggregations independently, and then combine the results in such a manner that the output from parallel processing is the same as if the processing happened sequentially. MOLAP operators in particular can be designed to be shared-nothing, since distributed aggregation is generally parallelizable.

3.5 Cube Data Initialization

TPC-DS data files are provided as a set of Hive tables. It's defined as bulk load process that implements a particular interface to inform the dispatcher node of several pieces of key information, which includes the data source and its schema, and also parse the queries to get the cube metadata.

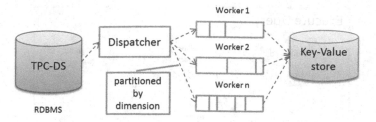

Fig. 5. Build the cube on workers

The bulk load process builds the cube by extracting data file records from Hive while also populating the multidimensional array (see Fig. 5). Records will be sent to multi-actors to be handled concurrently in a round robin fashion. This is generally the most time consuming part when loading large amounts of data.

To construct the cube, the dimension instances are generated first. Then the distinct members for each attribute are assigned bit keys, and stored into HBase as index.

After we partition tuples in fact tables into small parts to distributed workers according to the dimension chosen by user or by default. Let each node load tuples into multidimensional array. A tuple is represented as a cell in the multidimensional array indexed by the bit key of each of the attributes. Hence, each measure needs to be loaded in the multidimensional array from the tuples. Here we use the hash based method [14].

Finally, the engine performs aggregation calculations. Since we create the base cuboids first and then aggregation cuboids during user querying, all aggregation can be done in memory locally. For aggregation cuboids, the dispatcher node schedules the map task; each worker node sends the aggregated data to one reducer node that is appointed by the dispatcher. After base cuboids are loaded, it will be held in memory and can be updated and refreshed by the dispatcher.

3.6   OLAP Query Execution

When user OLAP query requests are submitted to the dispatch node, the node will parse and decide whether the query needs new cuboids or existing cuboids. If new cuboids are required, the dispatcher sends cuboid construction messages to associated worker nodes. Otherwise, it compiles the query into filter and aggregation expressions, and then sends them to each worker node.

Each worker node has a query executer that handles the message and sends back the hit cells to a node that accumulates all hit cells. The message includes the compiled filter that can be computed with key of cells in specified cuboid, and if the cuboid does not exist, the base cuboid will be used and aggregated into the queried cuboid. The node merges all results and transforms them from key-value pairs into records with dimension member, and returns them to the dispatcher.

As Fig. 6 shows that the query executer manages amounts of mappers and reducers for the extractor and filter. There are two kinds of applying filters: The first type scans all the records and check whether it passes filtering constraints. If it does, it is sent to reducers and taken into account. Reducer helps to merge and sort the result. The second

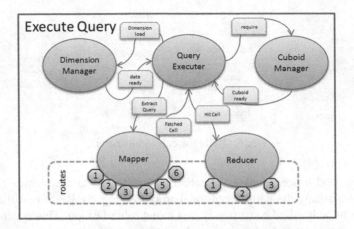

Fig. 6. Actor model for OLAP query executing

one tries to apply the filter to combine the dimensions that are not used in the filter. Thus each mapper will get an assembly key and try to fetch from HBase or memory cache. If it finds the keys, the mapper sends key-value pairs to the reducers. Here mappers and reducers can run on multi-threads, which gives the benefits to be concurrent locally and remotely. All actors in the engine communicate with each other by means of asynchronous messages that well define the partitions of data. Cube data cannot be shared between actors and be handled only once by each actor; the actors in different roles processes messages sequentially. The number of mappers and reducers can be configured according to system load.

3.7   Resilient Distributed Datasets (RDDs)

The OLAP engine, as shown in Fig. 7, is implemented with Resilient Distributed Datasets (RDDs) [16]. It performs most computations in memory while offering fine-grained fault tolerance. The engine will use RDDs to cache the base cuboid for the queried cube and produce aggregation cuboids when queries need them.

Firstly, we set an RDD as a read-only, partitioned multidimensional array. It can be created by the worker from local data of the base cuboid table.

Secondly, the worker has enough information for RDD constructing from the message received. Since workers are running on HBase region nodes, and it can load cuboid parts locally, it can materialize the RDD efficiently in any time when needed.

Finally, the worker controls the RDD's partitioning by pointing a key dimension and member range when other cuboids are aggregated from base cuboids.

The RDD just includes the base cuboid data and benefits OLAP operations that query other cuboids directly. Thus the cube data is only base cuboid size finally. We also can cache other cuboids dynamically for future iterations and we will research on the dynamical partition the other cuboids.

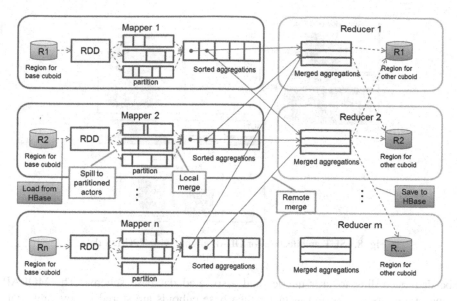

Fig. 7. The workflow of building other cuboids

4 Experiments for Big Data Benchmark

Big data analytics systems have many characteristics such as: scale-out for terabytes data, complex data types, emphasized user-defined functions, high fault-tolerance and require low latency [20]. Recently a few of big data component benchmarks, like Berkeley big data benchmark, Terasort etc., are emerged, but they are special for a subset of components benchmarking. We need an end-to-end benchmark to provide various test scenarios (workload) to verify most of above big data characteristics and abide following benchmark principles: self-scaling; results should be comparable at different scales, technology agnostic and simple to run [21].

OLAP is a typical and irreplaceable enabler of current business analytics systems so that future big data analytics systems have to provide the OLAP engine. TPC-DS, the latest decision support benchmark for relational databases, lacks some characters of big data like semi-structured and unstructured data and their associated analytics workload [22]. Therefore, BigBench based on TPC-DS is proposed as an industry standard benchmark for big data analytics [23]. Compared with other benchmarks experiments, TPC-DS benchmarking on big data systems can reuse the traditional benchmark knowledge and compare SQL and NoSQL solutions on most of key-value data stores. In advanced experiments, we can combine both SQL and NoSQL benchmarks based on the cube model of TPC-DS in Fig. 8 testing environment. In the following experiment, TPC-DS SQL are rewritten in MDX forms, which are decomposed into various kinds of implementations of cube algebra query operators (native API) [13].

Ref. to Fig. 8, the system under test (SUT) are composed of the NoSQL database its native API. MOLAP engine is taken as a core of test tool. We add a cube modeling module in the MOLAP engine to generate the cube metadata before populating the

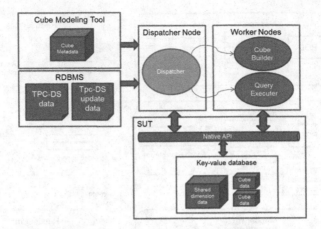

Fig. 8. SUT architecture for OLAP engine with TPC-DS data

cube data according to user queries and associated database schema. The MOLAP engine builds first dimension instance, then base cuboids are stored in key-value databases. As such, the MOLAP engine can accept user MDX analytics requests and execute cube algebra operations on key-value systems to provide an end-to-end-like OLAP benchmark testing scenario. The benchmarking procedure consists of following stages:

1. Generating TPC-DS datasets with different scale factors, and loading them into a relational database (Hive used in the following experiment).
2. Designing cube algebra operators and building the cube model according to TPC-DS query templates in the testing driver.
3. Generating base cuboids according to specified testing queries.
4. Executing user TPC-DS queries on the MOLAP server.

In the experiment, we reuse not only the TPC-DS data generator but also SQL engine on cloud systems to load TPC-DS data files to relational databases which can be used to test the performance of ROLAP on key-value systems. The SQL templates are transformed into MDX forms in order to find the dimensions and measures of building cube model in step 2. For example, from Query 7 we can find:

```
Dimensions:  "i_item_id",  "i_category",  "i_brand",
     "cd_gender",  "cd_marital_status",
     "cd_education_status",  "p_channel_email",
     "p_channel_event",  "d_year"
Measures: "ss_quantity", "ss_list_price",
     "ss_coupon_amt", "ss_sales_price"
```

An important question for the experimental study is the choice of input data. We use three sets of TPC-DS data in sizes of 1 G, 10 G and 100 G. The experiments utilized 3 nodes with 6 X Intel Xeon CPU E5-2640@ 2.50 GHz, 15000 r/s SAS hard disk and 256G memories. HBase and Hive are running with out-of-the-box setting on the Cloudera cluster included three nodes. The OS is Ubuntu Server 12.04. The network band width is 10 Gbps.

4.1 Cube Data Population Testing

During cube data population (step 3), only base cuboids are constructed in the key-value store from relational databases. We can measure the performance cube data generation from the following cube construction phases:

1. Initializing dimension members, execute query on TPC-DS dimension tables from RDBMS and store dimension members in the key-value store.
2. Querying table data, join TPC-DS fact table with above dimension tables and get all fact data.
3. Aggregating and sharding cuboids, send fact data partitioned according to dimension with maximum members to worker nodes and do aggregation computing in each worker nodes to construct related cuboids.
4. Saving cuboids, save all cells of cuboids into key-value database. In this stage, we can improve the performance by setting actors numbers for concurrent I/O benchmark under intensive read/write.

We measure the performance for each construction phase of cube data population on 1 G, 10 G, and 100 G of TPC-DS dataset and find the results are linear speedup at different scales (Fig. 9). This proves that the cube data building is suitable for scalability benchmarking.

This cube data population mode gives recipe to evaluate heterogeneous physical structures which have been implemented in current open NoSQL systems. For example, there are two approaches for cube data aggregation: compute in-memory in OLAP engine or call key-value database functions. The former needs much memory to hold the intermediate result. The latter push the aggregation calculation into database by native APIs if we want to benchmark key-values systems I/O operations. HBase, MongoDB, and Redis have pushed down functions to do summary and count aggregation that will help them keep comparable efficiency to other low level API systems.

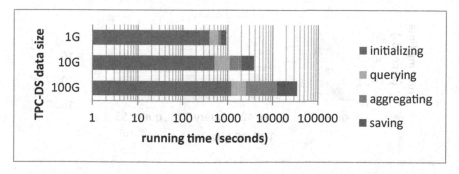

Fig. 9. Base cuboid building performance on 1 G, 10 G and 100 G TPC-DS data

4.2 Cube Querying

Cube querying in the MOLAP engine includes 4 stages: (1) loading base cuboids as RDD, (2) if needed, creating aggregation cuboids according to user query's summary expressions, (3) on worker nodes filtering with the compiled bit keys that are transformed from the query conditions, (4) worker nodes sending intermediate results to dispatcher node and the dispatcher node merging results.

Here we give the testing result of the base cuboids loading. RDD is built by reading local partition data and the MOLAP engine can create multiple actors to load base cuboids concurrently. Aggregation cuboids are computed from base cuboids, the partition of base cuboids makes the aggregated computation distributed and paralleled. Figure 10 shows the RDD loading performance at different scales. We find that sharing data via RDDs greatly speeds up next aggregation operation iterations.

Then we experiment other TPC-DS queries, a set of filter keys are calculated and applied on each partition. After filtering, the results are sent to dispatcher that merges all the results. After all workers send their results, the dispatcher renders results to convert bitwise key to dimension members and returns them to the client. We find the query result is returned in the same order we send them (Fig. 11).

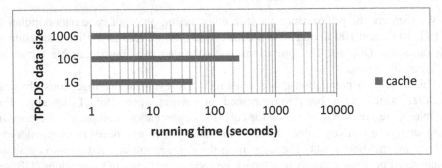

Fig. 10. RDD loading based on cube metadata from TPC-DS query 7,42,52,55

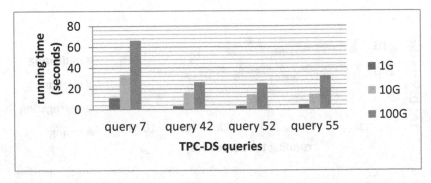

Fig. 11. Cube querying based on cube metadata from TPC-DS query 7,42,52,55

5 Conclusion

This paper presents a feasible way to build MOLAP engines on key-value stores for large-scale data analytics. It uses the partition of the HBase and parallel computations with actor programming model. All of them make the MOLAP engine support scale-out and fault tolerance.

Overall, the implementation has been successful in most of its original goals. Basic cuboids are constructed from involved queries for fact tables by using multidimensional array technology, and cuboids for various granularity aggregation data are derived with basic ones at running time by using distributed and parallel dynamic cube data load workers. Cuboids are transformed into linearized arrays which are replaced by bit key suitable for key-values data stores. The OLAP engine implemented with Resilient Distributed Datasets (RDDs) can perform most computations in memory and cache the aggregation data in memory for quickening the interactive responds on large-scale data while offering fine-grained fault tolerance. The features provided by the prototype enable fast computation of results for some TPC-DS queries. The update of cube data will be touched in future.

From experiment results, we found the cube building and cuboids caching technologies need to be improved in various ways. For example, it is better to use MapReduce functions to create basic cuboids from Hive tables and store them in the HBase. This will resolve the dispatcher's bottleneck found in the above experiments. Another key for performance is to use distributed high-dimension index for interactive OLAP queries.

In the future, different cluster parameter settings will be included into the experiments. We give only the out-of-the-box testing experience in this paper. During experiments we also evaluated for business question, I/O costs of MapReduce tasks, and the cost of data transfer for remote messages. By investigating such data, we envision the implementation of the MOLAP engine for handling most of big data analytics. As such we can abstract an interface that leverages other kinds of key-value stores since there are a lot of NoSQL database supports for cube data naturally.

Acknowledgements. This work was supported by the National Science and Technology Major Projects (2008ZX01045-004-01, 2009ZX01045-004-001 2009CB320706).

References

1. Cuzzocrea, A., Il-Yeol, S., Karen, C.D.: Analytics over large-scale multidimensional data: the big data revolution!. In: Proceedings of the ACM 14th International Workshop on Data Warehousing and OLAP. ACM (2011)
2. Evelson, B.: It's the dawning of the age Of BI DBMS. Technical report (2011)
3. Melnik, S., et al.: Dremel: interactive analysis of web-scale datasets. Proc. VLDB Endow. **3**(1–2), 330–339 (2010)
4. Corbett, J.C., et al.: Spanner: Google's globally-distributed database. In: Proceedings of 10th USENIX Symposium on Operating Systems Design and Implementation (2012)

5. Xin, R., et al.: Shark: SQL and rich analytics at scale. arXiv preprint arXiv:1211.6176 (2012)
6. Sanjay, G., Choudhary, A.: Sparse data storage of multi-dimensional data for OLAP and data mining. Technical report CPDC-TR-9801-005, Center for Parallel and Distributed Computing, Northwestern University (1997)
7. Turcu, A., Binoy R.: Hyflow2: A high performance distributed transactional memory framework in scala. Technical report, Virginia Tech (2012)
8. Duda, J.: Business intelligence and NoSQL databases. Inf. Syst. Manag. 1(1), 25–37 (2012)
9. Poess, M., Nambiar, R.O., Walrath, D.: Why you should run TPC-DS: a workload analysis. In: Proceeding of VLDB (2007)
10. Cheung, D., Zhou, B., Kao, B., Lu, H., Lam, T., Ting, H.: Requirement-based data cube schema design. In: Proceedings of the International Conference on Information and Knowledge Management. ACM (1999)
11. Niemi, T., Nummenmaa, J., Thanisch, P.: Constructing OLAP cubes based on queries. In: Proceeding of DOLAP (2001)
12. Niemi, T., Nummenmaa, J., Thanisch, P.: Constructing OLAP cubes based on queries. In: Proceedings of the ACM 4th International Workshop on Data Warehousing and OLAP. ACM (2001)
13. Ciferri, C., Ciferri, R., Gómez, L.I., et al.: Cube algebra: a generic user-centric model and query language for OLAP cubes. Int. J. Data Warehous. Min. (2012)
14. Goil, S., Alok, C.: High performance OLAP and data mining on parallel computers. Data Min. Knowl. Disc. 1(4), 391–417 (1997)
15. Zaharia, M., et al.: Resilient distributed datasets: a fault-tolerant abstraction for in-memory cluster computing. In: Proceedings of the 9th USENIX Conference on Networked Systems Design and Implementation. USENIX Association (2012)
16. Romero, O., Abelló, A.: Multidimensional design by examples. In: Tjoa, A.M, Trujillo, J. (eds.) DaWaK 2006. LNCS, vol. 4081, pp. 85–94. Springer, Heidelberg (2006)
17. Jianzhong, L., Rotem, D., Srivastava, J.: Aggregation algorithms for very large compressed data warehouses. In: Proceedings of the VLDB (1999)
18. Taylor, R.C.: An overview of the Hadoop / MapReduce / HBase framework and its current applications in bioinformatics. BMC Bioinform. 11(Suppl 12), S1 (2010)
19. Dean, J., Sanjay, G.: MapReduce: simplified data processing on large clusters. Commun. ACM 51(1), 107–113 (2008)
20. Alexandrov, A., Brücke, C., Markl, V.: Issues in big data testing and benchmarking. In: Proceedings of the Sixth International Workshop on Testing Database Systems. ACM (2013)
21. Baru, C., Bhandarkar, M., Nambiar, R., Poess, M., Rabl, T.: Setting the direction for big data benchmark standards. In: Nambiar, R., Poess, M. (eds.) TPCTC 2012. LNCS, vol. 7755, pp. 197–208. Springer, Heidelberg (2013)
22. Baru, C., Bhandarkar, M., Nambiar, R., et al.: Benchmarking big data systems and the bigdata Top100 List. Big Data 1(1), 60–64 (2013)
23. Ghazal, A., Hu, M., Rabl, T., Raab, F., Poess, M., Crolotte, A., Jacobsen, H.A.: BigBench: towards an industry standard benchmark for big data analytics. In: Proceedings of the SIGMOD (2013)
24. OLAP4J. http://www.olap4j.org/. Accessed 17 Feb 2014

Big Data Cloud-Based Advisory System

Vladimir Suvorov[✉]

EMC Corporation, Saint Petersburg, Russia
vladimir.suvorov@emc.com

Abstract. This article proposes a concept of Big Data Advisory & Benchmarking platform in a virtualized environment based on VMware vSphere. The key value of the proposed platform is a unified Big Data business-oriented benchmarking approach, data collection approach and an advisory system based on sequential process modeling and cost comparison model. Unified benchmarking ensures same tasks using the same data. Data collection utilizes vSphere metrics for system performance. A prototype solution is developed.

Keywords: Big data benchmark · Big data advisory system · Hadoop benchmark · Cassandra benchmark · Choosing big data technology · Graphchi benchmark · Virtualized big data · Data science

1 Introduction

Big Data now is driving the IT industry. Multiple platforms of various kinds emerge every now and then. All of them are still devoted to the same task – solve business problems, help sell something, provide predictive analytics. Before the Big Data concept most of the data was kept in relational databases and an SQL language was de facto standard for analytics.

Nowadays Big Data brings up challenge of dealing with Variety, Velocity, Volume and that becomes a challenge for IT professionals. A good Database administrator could relatively quickly switch from Oracle [1] to MSSQL [2] and the basic database design principles still hold. NoSQL approach does not have any rules. For example, Hadoop and Cassandra architectures are completely different, data storage concept is different, computational paradigms are different, preferred programming language is different. MongoDB [3] and GraphChi [4] have its own features. Also Twitter Storm can be mentioned as another approach, not to try to estimate difference between Hadoop-based extensions (Pivotal HAWQ, Cloudera Impala, Apache Drill, Hive, etc.). One needs a huge expertise to choose in this zoo.

There is no industry standard on benchmarking such systems. Each technology has its own tests, for example, Hadoop [5] has Teragen, Terasort and TeraValidate and DFSIO tests and there is the Yahoo Cloud Serving Benchmark [6] for database-like systems and many other specific benchmarks.

Many new emerging benchmarks are based on the same principles as the TPC's benchmarks [7] that regulate the data structure and queries to the data. For example, there is an upcoming initiative of BigData Top100 [8], which touches the question for a

© Springer International Publishing Switzerland 2014
T. Rabl et al. (Eds.): WBDB 2013, LNCS 8585, pp. 171–178, 2014.
DOI: 10.1007/978-3-319-10596-3_13

general case and recommends the benchmark types and the process. However, usually general benchmarks like TPC's benchmarks do not provide the reference implementation and these benchmarks are implemented mostly by vendors.

The scope of this research is focused on the technical part of creating a scalable platform for running benchmarks and getting performance rather than on specific data and queries. The platform that is proposed allows running any kind of benchmarks and is intended for users.

2 Approach

There are basically two approaches to benchmarking. Either the test dataset and the way operations are performed is strictly specified or they may vary to any extent. As an example of the benchmark of the first type one can consider 3DMark [9] test for video gaming performance. This test includes several popular games and several screen resolutions with fixed video options. Usually the system having higher score will perform better overall, but it is not guaranteed that your "World of Tanks" will run better on system A than on system B even if it has better 3DMark.

TPC tests are traditionally used for testing database performance and recently TPC has created Big Data Working Group [10] to adapt tests for Big Data systems – that is to specify dataset and payload. Many suggestions on the workload characteristics since the first WBDB 2012 conference were made [11, 12] and now TPCTC conference is officially adopting Big Data Track.

This approach to benchmarking has the advantage of having clear universal results and easy setup. Most time the customer who wants some system does not run the test himself he just looks at the results and sees that "system A has query time 10 ms and system B – 12 ms". The problem is that a customer should hope that his workload is reduced to that exact universal workload for his area of interest.

Another approach is to make some benchmarking system without specifying the benchmark itself. Then the problem of mapping model to reality is still present but the model can be fine-tuned much more than when there is a pre-defined dataset with pre-defined workload. The drawback is that it takes time to design a model based on customer's workload and to run tests on this exact model and real customer's dataset.

Different companies such as Cloudwick Labs have developed projects that facilitate use case research and testing [13]. In this paper, the general approach and architecture of building such facilitating systems will be described.

3 Architecture

3.1 Overview

This general solution is built to facilitate deployment, manageability, task running and analytics in heterogeneous environments such as public or private cloud or even a bunch of hardware. There are 3 principal criteria. Comparability means we must be able to compare the results on the same hardware. Scalability ensures that the platform should scale easily to get performance for different clusters. Interoperability allows to operate on any platform and should be able to adopt any programming language. To make it detailed we impose the following functional requirements for the solution:

- Load the data to the technology
- Run specific tasks on the selected technology
- Get and store the resulting performance
- Monitor cluster configuration & health
- Interact with the user once test is finished or error emerges
- Monitor system performance
- Store its own configuration
- Ensure that same tests are run with the same data and queries
- Visualize the result

Considering general criteria the virtualization seems very reasonable due to several aspects. First is comparability. We would like to be able to ensure that we run on the same resources and it needs to be done fast. Second is that using virtual resources, we can always follow the system design guidelines for the specific Big Data technology and create as many nodes and share resources as recommended. The third practical reason is that now many Big Data systems are operated within a cloud and often need to vary in size on demand and this is enabled in the cloud by virtualization. The only reason not to use virtualization is performance concern, but papers show that virtualization overhead is minimal [14]. Also virtualization allows to utilize centralized monitoring and management systems. Some technologies like Hadoop stack can have automatic deployment in a virtualized environment like the Serengeti project [15].

Considering interoperability requirement, it is achieved by separating client and server using common interface in between. A client that resides on the technology side should utilize fastest available libraries, which are written in different languages for different technologies, that is Java for Hadoop and C++ for Cuda. The server is communicating with client by implementing common protocol.

It is reasonable to transfer data using the best and fastest file transfer protocols and do not wrap up datasets within some general-purpose protocols. The API itself should utilize one of the most common protocols available on many languages to ease adding technologies to the system. It is good to have human-readable protocol for the debugging and logging purposes when it is not high loaded with heavy messages, which is a good use case for SOAP or other XML-based protocol (Fig. 1).

The platform contains the following parts:

- Infrastructure side – number of servers running ESX and dedicated storage under vSphere[16] commandment. VMware vSphere was chosen, because of its manageability and great system performance monitoring ability.
- Technology side – a number of Big Data VM's each contains some number of nodes of some Big Data technology. It might be a Hadoop or Cassandra cluster or even plain native environment on a single machine. Big Data clusters might be deployed manually or automatically, existing ones might be used as well.
- Controller side – a number of Controller VM's. Each VM contains one or more technology connectors. Technology connector can interact with the technology-specific Big Data cluster and Control Service VM. Technology connector is

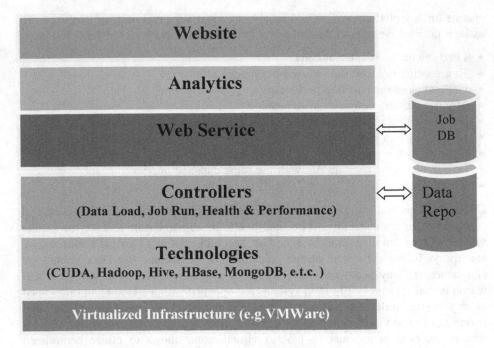

Fig. 1. General architecture

responsible for loading data and running external tasks for the technology. The task itself is not stored on the connector, instead it is externally loaded along with the data. Technology connectors can synchronize task execution with other connectors of the same type via SOAP messaging.

- Service side – Control Service VM + Test/Test Results Database + Data Storage. This is a command center which has CLI and GUI. It has its own database which stores configuration, test results, health history, also it stores paths to datasets and task samples. Task samples are technology-specific and mostly represent some library, object or executable file containing the sample task.
- Analytics module allows predicting results for unknown data and calculating cost-efficiency metric based on measured data and a number of preselected parameters. Different prediction and regression models are used.
- Website is merely an interface for the user.

3.2   Controllers

A Controller VM contains one or more Connectors. Connector is an instance that can communicate with Big Data Cluster for specific technology, (e.g. Hadoop Connector, Hive Connector) and implements specific SOAP interface for Service side. A Connector is platform and language independent and is only required to follow the interface. Some details are below (Fig. 2).

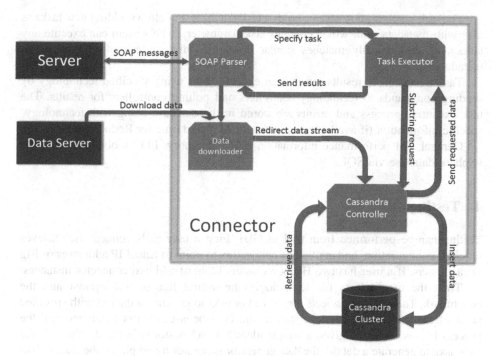

Fig. 2. Example - cassandra connector

The connector communicates with the server getting commands and sending the test results back. Commands can configure the connector for a specific cluster, tell the connector to take data from the data server and load it to a cluster and run/stop tests. Also the server can retrieve the connector's health, configuration, and test results. The connector utilizes FTP for data uploading and uses internal mechanism to ingest the data to the cluster.

Each connector has its own manifest, where it specifies the parameters it needs to self-configure. When a user adds the connector it sends back the list of parameters which are later configured in GUI or CLI.

3.3   Service Side

The service side includes a central database, which contains current test configuration, a task configuration service, a task execution & results service, and a data generator service. Service side has CLI and GUI service. GUI service connects to CLI part via SOAP and is a web-server utilizing AJAX and Java. GUI service is directly connected to database and can display a simple analytics over results.

The test configuration service holds configuration part, allows adding new connectors and monitors health of the system. CLI commands as well as a GUI are adaptively changed based on connector parameters.

The task configuration service holds a task database and allows adding new tasks as files with metadata xml, which contains task parameters. The system can execute any tasks and automatically matches similar tasks for different technologies based on metadata.

Task execution & results service can execute tasks using specified technology by sending commands to technology controllers and polling controllers for results. The task execution process and results are stored in the database along with technology-specific information (if available, e.g., time for Map and time for Reduce for Hadoop).

General VM performance information (CPU, Memory, I/O) is obtained from the vSphere database via SQL.

4 Testing

Testing can be performed from CLI or GUI. First a user adds connectors, retrieves required configuration, and configures connectors to point to actual IP addresses of Big Data clusters. If a user has two Hadoop clusters he must add two connector instances.

Then the user selects the technologies he would like to test against and the benchmark. Then the user selects the dataset or asks to generate a dataset with specified parameters. Then the user clicks start in the GUI or issues a run command and the process begins. A test is given a unique identifier and is stored in the database. If the user asks to generate a dataset the data generator generates it and places the dataset into the FTP server. The service side tells the connectors to download the data from the FTP server along with task files and specifies task parameters. When the connector downloads the dataset to a cluster it runs a task file against it and waits for the results. Each step is logged to the database. In parallel the health module on the service side monitors the connection state and technology-specific data (e.g. % of Maps and Reduces) and the results service collects VM performance information from vSphere. Upon success the results are returned from the connector to the results service. Upon fail the health module/config service will immediately report.

A user can view different graph plots and a comparison table along with all information for a particular test.

5 Advisory Service

An advisory service directly results from the benchmarking. The real business case is modeled with the sequential parameterized tests. As a sufficient amount of tests were run, a prediction model is applied, when a general model is based on known scalability of the technology and a user is given parameters like average execution speed of the pipeline and fault tolerance. As all technologies are based on a virtualized environment the performance data can show resource utilization, therefore an estimation of the cloud cost is available. This estimation is provided both for Amazon Cloud and private cloud. Also license pricing is available and average maintenance and developer cost. So platforms are competing based on money value, not poor performance. This approach utilized with module structure of connectors that require only a simple interface allows to compare even such platforms like Hadoop and CUDA.

6 Prototype

The presented approach was prototyped in a small cloud and had the following implemented connectors and cluster configuration (Table 1):

Table 1. Benchmarking environment

Technology	Number of nodes	Connector OS	Connector lang.	Task lang.
CPU	1	Linux	Java	C ++
Hadoop/Pig	4	Linux	Java	Pig
Hadoop	4	Linux	Java	Java
OpenMPI	10	Linux	Java	Python/C ++
GraphChi	4	Linux	C ++	C ++
CUDA	1	Linux	C ++	CUDA
MongoDb	4	Linux	Perl	Perl
Cassandra	4	Windows	C#	C#

Several simple scenarios were tested like processing email data and searching for a particular string and processing website data and building a reverse index. Also some mathematical computations like matrix operations were tested. All these tasks consisted of smaller simple subtasks. These subtasks were implemented manually and inserted to the task database. The results that were produced correspond with common knowledge, for example matrix operations were handled by CUDA nicely and Hadoop was good for searching large number of unstructured emails.

7 Conclusion

The main goal was to propose the reference design of a system, which makes it easier to produce custom business oriented benchmarks for different Big Data systems.

The approach described in this article allows to have highly flexible testing environment, also the advisory approach is recommended for business-driven decisions. It allows cloud providers and cloud users decide wisely on big data platform to use for particular case and also allows load balancing prediction for cloud providers.

References

1. Apache Hadoop Project. http://hadoop.apache.org
2. Yahoo Benchmark. http://research.yahoo.com/Web_Information_Management/YCSB
3. TPC. http://www.tpc.org

4. BigDataTop100 Project. http://www.bigdatatop100.org/
5. Baru, C., Bhandarkar, M., Nambiar, R., Poess, M., Rabl, T.: Setting the direction for big data benchmark standards. In: Nambiar, R., Poess, M. (eds.) TPCTC 2012. LNCS, vol. 7755, pp. 197–208. Springer, Heidelberg (2013)
6. http://www.tpc.org/tpcbd/default.asp
7. Chen, Y., Raab, F., Katz, R.: From TPC-C to big data benchmarks: a functional workload model. Technical report No UCB/EECS-2012-174
8. Jia, Z., Zhou, R., Zhu, C., Wang, L., Gao, W., Shi, Y., Zhan, J., Zhang, L.: The implications of diverse applications and scalable data sets in benchmarking big data systems. In: Rabl, T., Poess, M., Baru, C., Jacobsen, H.-A. (eds.) WBDB 2012. LNCS, vol. 8163, pp. 44–59. Springer, Heidelberg (2014)
9. Ankus- the Big Data Cluster Lab Genie for accelerating use case research & testing. http://cloudwicklabs.com/labs/ https://github.com/cloudwicklabs
10. Berkley Big Data Benchmark. https://amplab.cs.berkeley.edu/benchmark/
11. Ames, A.J., Abbey, R., Thompson, W.: Big Data Analytics Benchmarking SAS, R, and Mahout. SAS Institute Inc., Cary, NC, Technical paper, 6 May 2013
12. Buell, J.: A Benchmarking Case Study of Virtualized Hadoop Performance on VMware vSPhere 5, Technical White Paper (VMware) (2011). http://www.vmware.com/files/pdf/techpaper/VMW-Hadoop-Performance-vSphere5.pdf
13. http://www.projectserengeti.org/

MPP SQL Engines: Architectural Choices and Their Implications on Benchmarking

Ravi Chandran[1]([⊠]), K.T. Sridhar[2], and M.A. Sakkeer[2]

[1] XtremeData, Inc., Schaumburg, IL, USA
ravi@xtremedata.com
[2] XtremeData Technologies Ltd., Bangalore, India
{sridhar, sakkeer}@xtremedata.com

Abstract. One approach to big data benchmarking is to study the implications of the underlying computing platform, both hardware and software. The software stack for big data analytics comprises a suite of technology solutions including MPP SQL Engines. This paper provides an overview of the architectural choices in the underlying hardware infrastructure, the architectural choices in the design of parallel SQL Engines, and the implications of both on benchmarking. A benchmark suite developed at XtremeData for internal engineering use is also described.

Keywords: MPP SQL · Cloud object-store · Benchmark

1 Introduction

The inflection point in computing solutions, initiated by the convergence towards virtualized-x86-Linux hardware and public cloud offerings, has been accelerated by big data [1]. The rapid growth in big data volumes is orders-of-magnitude larger than the growth in the traditional enterprise operational data. This has created tremendous challenges, since no enterprise can afford to ignore the value of big data analysis. Big data has also been democratized: where it was once the domain of global enterprises, now businesses of all sizes have easy access to and can benefit from big data.

Data volumes are growing much faster than traditional IT budgets can possibly increase, so incremental evolution of legacy technologies cannot meet the demands. Market pressures have spawned a wide range of new technologies in recent years. Many of these technologies are narrowly focused on specific applications. The focus of this paper is on structured datasets, essentially the traditional ecosystem of Data Warehouse-Data Marts (DW-DM) extended to include new big data sources. Unstructured or loosely-structured data [2], such as text files, documents [3, 4], audio and video, have their own specialized solutions. The challenge of structured big data analytics is to deploy solutions that scale affordably in the context of loading, joining, aggregating, analyzing and reporting on billions of records quickly enough to meet business demands.

Over the past few years, the market has seen the rise of many "NoSQL" [5] solutions that are still evolving today; but the general consensus emerging is that Hadoop-like solutions need to be augmented by a full-featured SQL engine that can

© Springer International Publishing Switzerland 2014
T. Rabl et al. (Eds.): WBDB 2013, LNCS 8585, pp. 179–192, 2014.
DOI: 10.1007/978-3-319-10596-3_14

operate at large scale. SQL skills are ubiquitous in the market and a MPP (Massively Parallel Processing) SQL engine that hides the complexity of parallel execution is an attractive solution that can be immediately deployed.

In the context of benchmarking big data systems, it might be useful to study the architecture of the underlying infrastructure: both hardware and software. The focus of this paper is restricted to only one particular component of the big data software stack: a MPP SQL engine. In the following sections, the hardware infrastructure choices are first described, followed by the architectural choices for the MPP SQL engine. This is followed by a discussion of the implications on benchmarking.

2 Hardware Infrastructure

All components of a big data solution need to meet one common criterion; the ability to be deployed on today's converged, sharable hardware infrastructure. This converged architecture applies equally to public clouds, like Amazon Web Services (AWS) [6] and Google Compute Engine (GCE) [7], and to private clouds within the enterprise data center. Hardware infrastructure in today's converged data center comprises three layers: server-network-storage.

2.1 Server

By definition the server layer is a virtualized-x86-Linux machine; other CPUs and operating systems are not significant contenders. Given this, there is very little differentiation in terms of hardware performance or functionality between vendors of x86 machines: all are more or less equal.

2.2 Network

The network layer offers much more interesting choices. Unlike the server layer, the performance of the network tends to be more of a step function rather than continuous: 1 or 10 gigE, DDR[1] or QDR[2] InfiniBand. Network switches also have a range of capabilities that can significantly impact data-intensive processing: bi-section bandwidth, lane throttling/bandwidth guarantees [8]. Today the most prevalent network infrastructure is based on 10 gigE.

2.3 Storage

The storage configuration for data-intensive processing has diverged away from the traditional SAN[3]/NAS[4] and has converged to two choices: "object store" and

---

[1] Double Data Rate.
[2] Quad Data Rate.
[3] Storage Area Network.
[4] Network Attached Storage.

"distributed filesystem". This shift was driven by the need for scalability in MPP systems, which shared solutions like SAN/NAS cannot provide. The new offerings in storage are exemplified by OpenStack [9]: Cinder (block store, filesystem) and Swift (object store). These offerings are very similar to those offered by public cloud providers, like AWS and GCE. Underlying both these solutions is the same hardware infrastructure: a cluster of machines with locally-attached storage. This cluster is typically built with the same x86-Linux machines user in the server layer, but other choices may be also be employed in the future, such as low-power ARM CPUs. The storage medium itself uses rotating magnetic disks, flash memory or some combination.

The block store could be local storage attached to each Linux server or a distributed solution spread across multiple servers. The object store is a scalable redundant storage system that offers the features of a traditional SAN/NAS system, but in a much more scalable fashion, leveraging commodity Linux clusters.

This combination of object and block store enables a re-definition of the traditional Data Warehouse architecture. Figure 1 shows a simplified block diagram of such a traditional DW, comprising two major components: a reliable storage layer and an SQL querying layer. In today's cloud architecture, storage can be decoupled from the SQL querying engine, as shown in Fig. 2. The reliable storage layer is provided by the cloud object-store. The SQL querying layer is configured as a set of computing resources in the cloud, populated with data from the object store. The populated data can reside either in a cloud block store or locally in storage associated with the computing resources, during the lifetime of the SQL querying engine. When the engine is shutdown, changed data can be uploaded back to object store, and re-populated on a re-start.

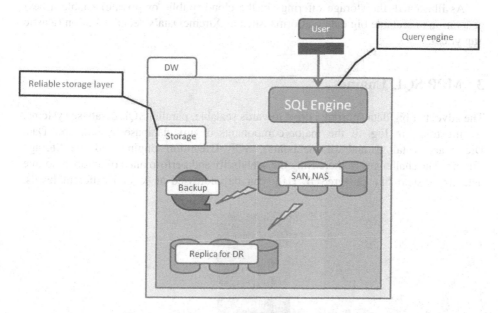

Fig. 1. Architecture of traditional Data Warehouse

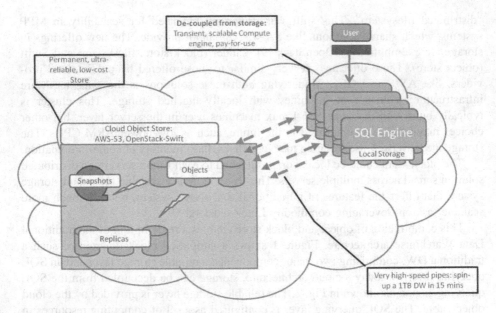

Fig. 2. New Data Warehouse architecture enabled by the Cloud

This decoupling enables the storage and querying layers to scale independent of each other. This is especially so, if the SQL querying engine is designed to be parallel, shared-nothing in its architecture (MPP), as discussed in the next section.

As illustrated, the storage offerings in the cloud (public or private) enable a new generation of scalable big data platforms, such as XtremeData's "elan" solution (**el**astic **an**alytics).

3   MPP SQL Engines

The advent of big data spurred a drive towards scalable, parallel SQL database systems. As illustrated in Fig. 3, the major components of any database system are: Data Dictionary (catalogs and all meta-data), SQL Execution Engine and the Storage Engine. The challenges to increasing the scalability and performance of an open-source database system like postgreSQL [10], can be addressed at several different levels,

Fig. 3. Components of a DB Engine like postgreSQL: starting point for MPP SQL

leading to implementations that may be characterized as "Sharded", "Federated" or "True-MPP", as described below.

Early implementations were "Sharded": tables were broken up into shards and each shard was stored in a separate SQL database instance, typically an open-source engine like MySQL and postgreSQL. Later, commercial vendors incorporated the sharding process in their offering, leading to "Federated" solutions, where the end-user did not have to directly handle the sharding, as shown in Fig. 4.

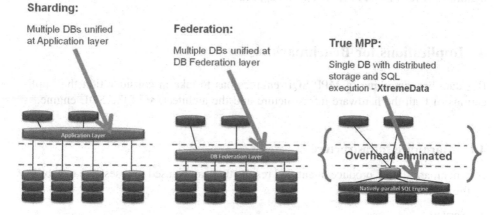

The higher the level at which parallelism is implemented, the higher the overhead: more hardware required to do the same job.

Fig. 4. Architectural approaches to MPP SQL

Almost all commercial offerings in the market today are federated systems, with multiple complete instances of postgreSQL running under the hood. From a benchmarking viewpoint, this is worth examining a little more closely, since the implications of multiple complete instances are very significant.

Within postgreSQL, the core SQL execution engine (although evolving continuously) is single-threaded. This has a direct impact on the usage efficiency of the underlying hardware, since the x86 CPU's path is in the direction of many cores. A single-thread can only use one core. Therefore to leverage the many-core CPUs (2x8-core CPUs per server is available today, more in the future), vendors of federated solutions recommend one instance of postgreSQL per core. This recommendation was reasonable even a few years ago, when there were only 2 or 4 cores per server. But with 16 cores today and probably 32 or 64 within a year or two, this recommendation rapidly degrades into unmanageable complexity.

A standard data center rack (42U) can hold 16 servers: this implies 256/512/1024 cores per rack: 256/512/1024 complete independent instances of postgreSQL in each rack is not a reasonable solution. This is especially so in the context of big data analytics, where joins and group-aggregates across multiple large tables are routine operations. These operations necessarily (in the general case, where data subsets cannot be guaranteed to be co-located in each node) require N:N exchange of data between all N nodes in the cluster. 'N' in this case refers to the number of logical nodes or database instances (256/512/1024) rather than physical servers (16 per rack). This is a serious limitation of federated designs, and any big data benchmark should be capable of quantifying the impact on performance and scalability.

A more manageable architecture is to limit the number of logical nodes, and create a natively-parallel, multi-threaded core SQL execution engine. A few vendors, including XtremeData, have taken this approach.

4 Implications for Benchmarking

Big data benchmarking of MPP SQL engines has to take in consideration the implications of both the hardware infrastructure and the architecture of the SQL engine.

4.1 Hardware Infrastructure

A benchmark should produce quantified results that can be used to assess the capability of the underlying hardware:

- Servers:
  - CPU capability (# of Cores)
  - Memory size and bandwidth
  - Network bandwidth

- Storage:
  - # of tiers (Memory, Flash, Disk)
  - size and bandwidth for each tier

- Network:
  - Topology: point-to-point latency
  - Switch bisectional bandwidth

Many tests are available in the industry to assess each of these hardware parameters independent of big data workloads. The results of these tests should also be reflected in a big data benchmark. For example, if CPU capability is poor (low clock frequency, small number of cores), the big data benchmark suite should include tests that clearly highlight this fact. In the benchmark we developed (described in later section), there are a series of queries that perform a full table scan of a single table, but with increasing complexity in the associated computation. This sets a baseline time for full table scan (typically limited by storage bandwidth) and assesses CPU capability.

4.2   SQL Engine

Assessing the SQL engine's capabilities for big data analytics requires tests that span these operations:

- Functionality:
  - SQL language support
  - Partitions, Indexes, Cursors, Window Functions
- Performance:
  - Parallel load from external source
  - Single table tests:
    - Scan-Filter-Complex compute
    - Group-Aggregate
    - Window Functions
  - Multi-table tests:
    - Joins
    - Joins + all of the above
  - Table creation within DB (Create Table As): for data-intensive, iterative processing
- Scalability:
  - Scale DB size while holding system size constant
  - Scale system size holding DB size constant

Assuming that a solution satisfies the functionality criteria, the keys metrics for big data analytics are Performance and Scalability.

Performance: This is obviously a key factor in the context of big data, when dealing with 100's of TBs deployed on clusters of 100's of x86 nodes. Physical footprint translates directly to costs of operation: every cubic-foot represents real-estate, power, cooling and labor costs. Minimizing physical footprint by extracting maximum performance from the underlying infrastructure is a fundamental requirement.

Scalability: An ideal system will support linear scalability across two dimensions: size of database and size of physical system.

A benchmark suite that can quantify metrics related to performance and scalability of SQL databases will be a useful component of any broader big data benchmark. The Transaction Processing Performance Council publishes a suite of benchmarks covering workloads that include On-Line Transaction Processing (OLTP) and Decision Support systems [11]. These benchmarks are domain-specific and try to simulate real-life workloads in order to compare systems from different vendors. Others have approached benchmarking from different perspectives [12, 13].

The next section describes a benchmark created specifically to simulate big data analytic workloads, from an engineering viewpoint rather than a business application viewpoint.

5   Benchmark

The design goals for the benchmark suite were: ease of use, portability across SQL platforms, and measurements across a range of metrics that would quantify performance and scalability – of both the SQL software layer and the underlying hardware layer.

In order to meet the goals of ease of use and portability, the benchmark is written entirely in SQL. This includes the code for table definition, data generation and table population, and a set of approximately 40 queries, organized into 8 groups. The entire SQL script is designed to be run with a simple command on the system under test, with the only parameters defining the range of scale factors to be tested. The benchmark creates tables that are large relative to system memory size, to ensure that storage systems are exercised. The benchmark also includes multiple large tables, to ensure that in a parallel system data is distributed across nodes, and the underlying network used for data movement is also exercised.

The benchmark script creates tables with columns that include integer, string and floating-point data types of various sizes (Table 1):

Table 1. Benchmark: column data types and quantity

Datum	Data type	Size (bytes)	Quantity
64 bits integer	bigint	8	7
32 bits integer	integer	4	15
String	varchar	8, 10, 12, 20, 34	17
float 8	double precision	8	5

Tables are created starting with the smallest at 32 K rows, with increasing sizes going up to 16 B rows. Each new table created is twice as big as the previous table.

Typically the benchmark is run at scale factors of 0 to 5, where the database size doubles for each scale factor increase. At scale factor 0, the database size is 0.33 TB and the largest table has 0.5 B rows, at scale factor 1, the database size is 0.66 TB and the largest table has 1 B rows. At scale factor 5, the database size is 10 TB and the largest table has 16 B rows. The data is generated in such a manner that the number of rows at each stage of the processing pipeline (Scan-Filter-Join-Group-Aggregate-...), approximately doubles with each increase in scale factor. Therefore it is a reasonable expectation that query execution times approximately double with each increase in scale factor. The queries are organized into 8 groups: Scan-Filter, Equi-joins, Non equi-joins, Group/Aggregation, Sub-Queries, Set Operations, Top N, Table Creation (Tables 2, 3, 4):

Table 2. Benchmark: scan-filter queries

Class	# of Tables	Output
Scan-Filter	1	Scan with filter of 0 % selectivity
	1	Scan with filter of 1 % selectivity
	1	Scan with filter of 10 % selectivity
	1	Scan with complex filer using bit-shift, double multiplication and division, 0 %
	1	Scan with complex CASE filter, 0 %
	1	Scan with CASE projection, sort with top ten rows

Table 3. Benchmark: join queries

Class	# of Tables	Output
Equi-Join	2	2-Table inner equi-join, 0 rows result
	3	3-Table inner equi-join, 0 rows result
	4	4-Table inner equi-join, 0 rows result
	2	2-Table inner equi-join with expressions, 0 rows result
	3	3-Table inner equi-join with expressions, 0 rows result
	4	4-Table inner equi-join with expressions, 0 rows result
	2	2-Table left outer equi-join, 0 rows result
	2	2-Table right outer equi-join, 0 rows result
	2	2-Table full outer equi-join, 0 rows result
	3	Inner equi-join on 3 sub-query selects
	5	5-Table inner equi-join; big/small tables; grouping; aggregation; sort; 10 rows result
	5	5-Table full outer equi-join; big/small tables; grouping; aggregation; sort; 10 rows result
Non Equi-Join	2	Nested loop: inner non equi-join; 0 rows result
	2	Nested loop: full outer non equi-join; 0 rows result
	2	Cartesian product; grouping; sort; limited to 10 rows result

6   Benchmark Results

The benchmark described above has been used to assess many different hardware platforms and several different SQL database engines, including XtremeData's dbX and a few leading competitors. In this section, some generalized results are presented, characterizing hardware platform layers and SQL software layers.

6.1   Hardware Platforms

The benchmark has been characterized on a variety of different hardware platforms, using dbX as the SQL engine. The key factors of the hardware platform that determine performance are: computing power, storage bandwidth and network bandwidth.

Table 4. Benchmark: group-aggregates, set operation and table creation queries

Class	# of Tables	Output
Group/Aggregate	1	Group by 4 cols; sort; limited to 10 rows
	1	Count on distinct rows sub-query; 1 row
	1	Count on distinct rows sub-query with expression; 1 row
	1	Group with aggregation; sort; limited to 10 rows
	2	Count on join sub-query producing distinct rows; 1 row
Sub-query	2	Count with filter using sub-query; 1 row
	4	Sub-queries in projection and filter; 1 row
	2	Union; limited 10 10 rows
	2	Intersection; 0 rows
	2	Except: limited to 10 rows
Top Few	1	10 unique rows of a full table scan
	1	10 unique rows of a group sub-query
	2	10 unique rows of a join sub-query
Table Creation	2	CREATE TABLE AS: join sub-query result
	2	INSERT INTO: join sub-query result (duplicate rows)
	2	SELECT INTO: join sub-query result

As a rough measure of computing power, we use "Ghz" as the metric. For example, a 4-core 3.0 Ghz Xeon will be rated as 12 Ghz. This is not a perfect metric by any means, but is reasonable for these purposes. Storage and network bandwidth are directly measured in GB/s.

	XD-1	XD-2	VH_SSD	VH_SAN	XD-8	VC-1	VC-2
CPU	Nehalem	Xeon	Xeon	Xeon	Opteron	Nehalem	Nehalem
Storage	Direct	Direct	SSD	SAN	Direct	Direct	Direct
Network	IB	IB	Fiber	FC	IB	10GigE	10GigE

Where:

- XD-1/2/8: Refers to XtremData configured platforms of 1/2/8 nodes. Each node is a dual-socket Opteron server with direct-attached disks and a DDR 4xInfiniBand (IB) network.
- VH-SSD/SAN: VendorH, a Tier-1 enterprise-grade system with an SSD fiber-attached array (custom link) or a FiberChannel (FC) attached SAN.
- VC-1/2: VendorC, a Tier-2 enterprise-grade system with dual-socket Nehalem servers, direct-attached disks and a 10 gigE network fabric.

The relative strengths of CPU power across platforms (Table 5):
The relative strengths of Disk bandwidth across platforms (Table 6):
Since these hardware platforms cover a broad spectrum of physical footprint and costs, it is difficult to make direct performance comparisons. But it is instructive to isolate queries that are known to be CPU-limited (heavy computation) and queries that

Table 5. Hardware platforms: CPU capability

	XD-1	XD-2	VH_SSD	VH_SAN	XD-8	VC-1	VC-2
CPU type	Nehalem	Xeon	Xeon	Xeon	Opteron	Nehalem	Nehalem
Clock, Ghz	3.3	2.4	2.26	2.26	2.4	2.9	2.93
# Cores	6	6	8	8	4	8	6
# Sockets	2	4	8	8	16	8	16
Total, Ghz	39.60	57.60	144.64	144.64	153.60	185.60	281.28

Table 6. Hardware platforms: Disk bandwidth

	XD-1	XD-2	VH_SSD	VH_SAN	XD-8	VC-1	VC-2
Disk, GB/s	1.60	2.0	4.00	5.00	5.20	8.36	10.64

are known to be I/O limited (heavy data movement), and compare relative performance.

CPU-Limited Queries: Two queries were selected (c.2 and c.5), both performing DISTINCT operations with Joins and sub-queries. As shown, the performance of the CPU-limited queries correlates well with the CPU strengths of the hardware platforms (Fig. 5).

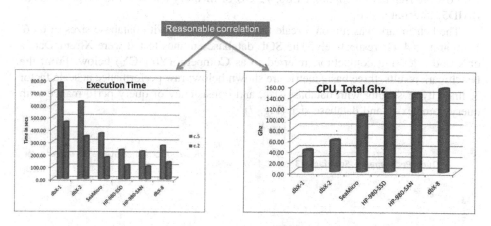

Fig. 5. Benchmark Results: CPU-limited queries

I/O-Limited Queries: Three queries were selected (b.1, b.2 and b.3), performing 2, 3 and 4 table Joins respectively. Again, as shown, the performance of the I/O-limited queries correlates well with the Disk bandwidths of the hardware platforms (Fig. 6).

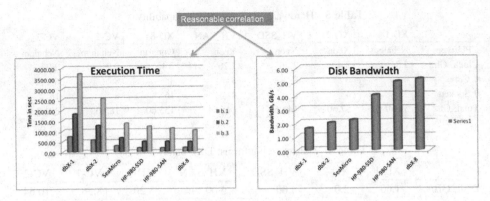

Fig. 6. Benchmark Results: I/O-limited queries

6.2   SQL Engines

Another point of reference on the usefulness of the benchmark is to characterize the performance of different SQL database engines running on the same underlying hardware infrastructure. This exercise equalizes the hardware and highlights the strengths and weaknesses of the SQL software layer.

The results below were performed on a single physical node: two-socket server with 6-Core Nehalem 3.33 Ghz CPUs, 72 GB of memory and 6 × 15000 rpm disks as RAID5, direct-attached.

The benchmark was run on 3 scale factors: 1, 2 and 3, with database sizes of 0.66, 1.32, and 2.64 TB respectively. The SQL database engines tested were XtremeData's dbX and a leading competitor, referred to as CompetitorY(or CY) below. From the benchmark results, three basic metric are shown below: raw performance at scale factor 3, linearity of scaling with database size, and consistency of query performance with query complexity and database size (Fig. 7).

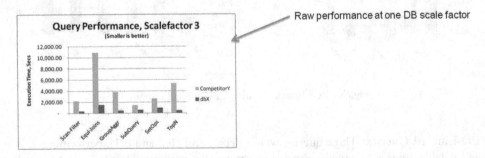

Fig. 7. Results: SQL database engines: raw performance

Raw query performance at the fixed scale factor of 3 (2.64 TB), shows that dbX was consistently faster across all query groups, an average of 7–10x faster.

Linearity of scaling and consistency of performance is illustrated next. In both charts, the vertical axis is of the same scale, and denotes normalized query execution times, with scale factor 1 set to 1. Therefore normalized execution times for scale factors 2 and 3 should ideally be 2 and 4 respectively – shown above as the "Ideal Linear" line in blue. Large divergence from the Ideal Linear line indicates non-linear scaling with database size. The charts show that dbX performance closely matches the Ideal Linear line, and also shows consistent performance across all query groups. CompetitorY results reveal significant sub-linear scaling with database size and also a large variance in performance across query groups (Fig. 8).

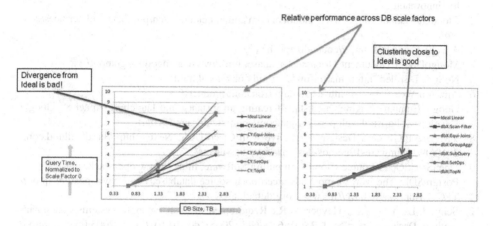

Fig. 8. Results: SQL database engines: scaling

7 Conclusions

Analysis platforms for (structured) big data need to include a full-featured MPP SQL database engine. The SQL engine should be capable of deployment in the shared, virtualized hardware infrastructure that today's data centers are converging towards, whether public or private Clouds. The SQL engine should offer high performance (maximize hardware efficiency) and offer near linear scalability with increases in database size and system size. These features determine physical footprint of the underlying hardware, which directly determines the total cost of operation. A benchmark has been developed to characterize a big data analysis platform: hardware resource layers and the SQL database layer. This benchmark is merely a starting point, and has several deficiencies:

- Synthetic data with known, fixed distribution
- All tables have same column schema
- Each new table is 2x previous table
- SQL data generation (INSERT) – can be slow!

Results have been presented using the benchmark across a variety of hardware platforms and on two competing SQL database engines. The benchmark is being continually updated to reflect newer features and application demands. Currently work is in progress on adding the following tests to the suite: ANI SQL Window functions, Insert/Update/Delete queries, continuous ingest during querying.

References

1. Big data: The next frontier for innovation, competition, and productivity, McKinsey& Company. http://www.mckinsey.com/insights/business_technology/big_data_the_next_frontier_for_innovation
2. Cloudera-Hadoop. http://www.cloudera.com/content/cloudera/en/products-and-services/cdh.html
3. Apache CouchDB. http://couchdb.apache.org
4. MongoDB: A Document Oriented Database. http://www.mongodb.org/about/
5. NoSQL Distilled. http://martinfowler.com/books/nosql.html
6. Amazon Web Services. https://aws.amazon.com/products/
7. Google Compute Engine:Quickstart: Creating an instance and launching Apache - Google Developers. https://developers.google.com/compute/docs/quickstart
8. VMware 10GE QoS Design Deep Dive with Cisco UCS, Nexus. http://bradhedlund.com/2010/09/15/vmware-10ge-qos-designs-cisco-ucs-nexus/
9. OpenStack Open Source Cloud Computing Software. https://www.openstack.org/
10. PostgreSQL: The world's most advanced open source database. http://www.postgresql.org/
11. TPC – Homepage. http://www.tpc.org/default.asp
12. Seng, J.-L., Yao, S.B., Hevner, A.R.: Requirements-driven database systems benchmark method. Decis. Support Syst. **38**(4), 629–648 (2005). doi:10.1016/j.dss.2003.06.002, http://dx.doi.org/10.1016/j.dss.2003.06.002
13. Darmont, J., Bentayeb, F., Boussaid, O.: Benchmarking data warehouses. Int. J. Bus. Intell. Data Min. **2**(1), 79–104 (2007). doi:10.1504/IJBIDM.2007.012947, http://dx.doi.org/10.1504/IJBIDM.2007.012947

Towards an Industry Standard
for Benchmarking Big Data Systems

Nambiar Raghunath[⊠]

Cisco Systems, Inc., 275 East Tasman Drive, San Jose, CA 95134, USA
rnambiar@cisco.com

Abstract. Big Data is one of most talked about topics across major industry verticals, research and, governments. Many enterprises are adapting Big Data analytics for improving operational efficiency, customer experience, and to drive new business opportunities. But they are challenged with tools, techniques and metrics to characterize and compare hardware and software systems that can optimally meet their requirements. In this paper the author looks at the industry treads, the role that well established industry standard consortia like the TPC can play in developing standards, and a status update from the TPC.

1 Introduction

Industry standard benchmarks have played, and continue to play, a crucial role in the advancement of the computing industry. Demands for them have existed since buyers were first confronted with the choice between purchasing one system over another. Over the years, industry standard benchmarks have proven critical to both buyers and vendors: buyers use benchmark results when evaluating new systems in terms of performance, price/performance and energy efficiency, while vendors use benchmarks to demonstrate competitiveness of their products and to monitor release-to-release progress of their products under development. Historically we have seen that industry standard benchmarks enable healthy competition that results in product improvements and the evolution of brand new technologies [1, 2].

Over the past quarter-century, industry standard bodies like the Transaction Processing Performance Council (TPC) has developed several industry standards for performance benchmarking, which have been a significant driving force behind the development of faster, less expensive, and/or more energy efficient system configurations [1–3].

The world has been in the midst of an extraordinary information explosion over the past decade, punctuated by rapid growth in the use of the Internet and the number of connected devices worldwide. Today, we're seeing a rate of change faster than at any point throughout history, and both enterprise application data and machine generated data, known as Big Data, continue to grow exponentially [1, 2]. It has been emerging as an integral part of enterprise IT across major industry verticals, research and, governments. A survey by Gartner, Inc reveals that Big Data investments continue to rise and 64 of organizations have invested in or plan to Invest in Big Data. 30 % have already invested in big data technology, 19 % plan to invest within the next year, and

© Springer International Publishing Switzerland 2014
T. Rabl et al. (Eds.): WBDB 2013, LNCS 8585, pp. 193–201, 2014.
DOI: 10.1007/978-3-319-10596-3_15

an additional 15 % plan to invest within two years, across major industry verticals have invested or plan to invest in Big Data [15]. See Fig. 1. These industry verticals represent current industry users of industry standard benchmarks from TPC and others.

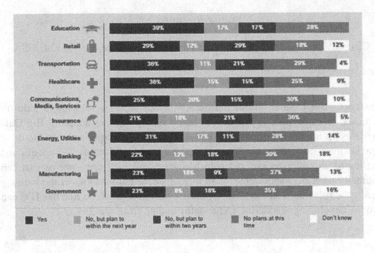

Fig. 1. Big Data investments across major industry verticals

The Big Data market is estimated to be at 17 billion in 2014, growing 30–40 % every year, and reaching 32 billion in 2017, and to $80 billion in 2020 [4]. See Fig. 1 that shows the Big Data revenue forecast from 2011–2017 between infrastructure, software and services [4] (Fig. 2).

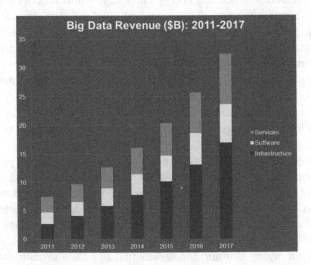

Fig. 2. Big Data revenue forecast

While enterprises realize the potential of Big Data, they are challenged with tools, techniques and metrics to characterize and compare hardware and software systems that can optimally meet their requirements. Due to lack of industry standards, vendors have been running their own benchmarks and making claims that are not easily verifiable by the customers. The situation is not a lot different from the 1980s, when several companies made their own performance claims on transaction processing capabilities of systems, that brought bunch of industry visionaries together to form the Transaction Percussing Performance Council [3] which laid the foundation for industry standards for benchmarking transaction processing systems, and later developed standards for decision support systems, energy efficiency data virtualization and data integration – all in line with trends in the industry [6].

2 Why TPC?

The TPC is a non-profit corporation founded to define vendor-neutral transaction processing benchmarks and to disseminate objective, verifiable performance data to the industry. Over the past quarter century, the TPC has had a significant impact on the industry and expectations around benchmarks. TPC benchmarks have permanently raised the bar; vendors and end users rely on TPC benchmarks to provide real-world data that is backed by a stringent and independent review process [3]. The main reasons why the TPC should develop standards for benchmarking Big Data systems are listed below.

- **In line with industry trend.** Historically, the TPC benchmark standards have been developed in line with the industry trends and end-user demands. Originally established to develop benchmark standards for transaction processing, later developed benchmark standards for decision support systems, virtualization and data integration in line with industry demands.

  The first benchmark TPC-A was evolved into TPC- TPC-B, and was replaced by TPC-C a 3-Teir Online Transaction Processing (OLTP) benchmark. TPC-C has been a standard since 1994 and doubtfully the post popular database benchmark tracking Moore's populations for enterprise applications [11]. See Fig. 3. TPC later added a new OLTP benchmark, TPC-E representing more complex transactions and system availability requirements which currently coexists with TPC-C.

  The first Decision Support benchmark from the TPC was TPC-D, which later evolved into TPC-H and TPC-R. TPC-R was retired due to lack of industry traction. In 2010 the TPC added a new DSS benchmark, the TPC-DS representing modern decision support systems with multiple business channels and large number of complex queries [12, 13]. While TPC-DS is yet to gain traction from a benchmark publication perceptive, it has been a very popular workload in the academia and industry as base for several workloads. Interestingly, there are few NoSQL (Hive etc) based implementations of it to measure performance of Big Data Systems.

  The TPC has also developed benchmarks for database workloads in virtualized environment and data integration to address industry demands.

  The next revolution in the data management platform space is Big Data and TPC is most well positioned to develop standards for it.

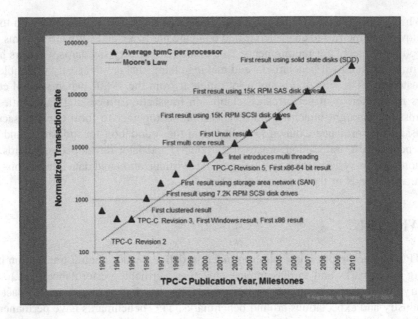

Fig. 3. TPC-C treads vs. Moore's Law

- **Established Organization.** At the helm of the TPC organization is the General Council, which is composed of all member companies. The TPC organization adheres strictly to the TPC Bylaws [20] and the TPC Policies [20]. The TPC organization is listed below:

  - **General Council:** All major decisions are made by the General Council through a democratic process. Each member company of the TPC has one vote.
  - **Steering Committee:** A five committee elected annually from member companies. The Steering Committee is responsible for overseeing TPC administration and overall direction and recommendations to the General Council.
  - **Technical Advisory Board:** This subcommittee is responsible for interpretation and compliance of TPC specifications and makes necessary recommendations to the Council.
  - **Public Relations Committee**: This subcommittee is tasked with promoting the mission and vision of the TPC.
  - **Technical Subcommittees**: There are two committee types. *Development subcommittee* is the working forum within the TPC for development of a Specification. *Maintenance subcommittee* is the working forum within the TPC for developing and recommending changes to an approved TPC Benchmark Standard.

- **Independent auditing process.** Verifiability is key requirement of any benchmark standard. There are three types of benchmark audits practiced in the industry - self-audit, peer audit and third-party audit, The TPC is has the most credible audit process combing all three. All TPC results are audited by independent auditors before they are published. The auditors go through a strict auditor certification process administered but elected committee to maintain high standards. After a benchmark result is published the TPC allows for 60 days every member in the TPC for peer-review during which any member can challenge published result based on technical correctness, pricing and system availability [3].

- **Well established documentation process.** All TPC benchmarks require an executive summary and a full disclosure report for every benchmark results available publicly. The full disclosure report documents the components, system under test, benchmark procedures and the source code for the programs used to enable test replication by any interested party. This full disclosure makes it possible to question and challenge a result and ensures that all published results are credible and verifiable [3].

- **Objective means of comparing price and price/performance.** TPC has the credibility of developing the most respected standard means of comparing the price and price/performance of different systems. As price/performance metric is a impacted by the configuration, the benchmark sponsors are pick realistic configurations in their benchmarks. There are several factors make TPC's pricing specification unique and creditable. TPC mandates that the pricing use in TPC benchmarks are verifiable and based upon a pricing model that the sponsoring company actually employs with customers. Due to this the TPC pricing is widely used by the end users in key purchase decisions [5].

  The pricing specification that describes the methodologies and metrics are common across all benchmark standards. Extending the well-established pricing specification to a new workload is fairly straightforward.

- **Objective means of comparing energy efficiency.** The TPC energy specification describes the methodologies and metrics to measure and report energy efficiency across benchmark standards. With the TPC-Energy metric, customers can identify systems that meet their price, performance and energy requirements via the TPC Web site. Extending this to a new workload is fairly straightforward as well.

  To help TPC benchmark sponsors reduce costs, the TPC provides a software suite, the Energy Measurement System (EMS). The EMS provides services such as power instrumentation interfacing, power and temperature logging and report generation. Even though reporting energy metrics is optional, competitive demands are expected to encourage vendors to include them. While energy efficiency is one of the important factors in evaluating Big Data systems, the well-established TPC Energy standard can have an important role [6].

- **Complete system evaluation.** The TPC benchmarking model has been the most successful in modeling and benchmarking a complete end-to-end business computing environment rather than subsystem evaluation. This has helped TPC benchmarks gain recognition as credible, realistic workloads. Most past and many current benchmarks only measure the hardware performance (processor and IO subsystems). TPC benchmarks have led the way in developing a benchmark model that most fully incorporates robust software testing [3] (Fig. 4).

- **Staying relevant.** The TPC has severed the industry well over past quarter century providing relevant and up-to-date benchmark standards. As Big Data becomes integral part of enterprise IT ecosystem across all major industry verticals, it is important for the TPC to continue to serve the industry with standards of relevance for its users. See Fig. 3.

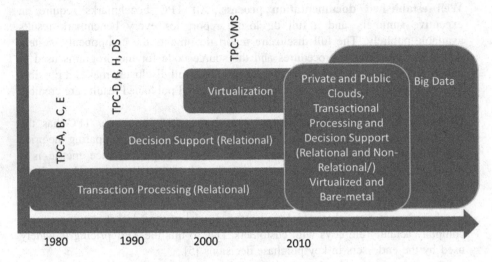

Fig. 4. Industry trends and TPC benchmarks standards

3 What does it Take?

Developing an industry standard benchmark is not and east task. It will take co-operation and handwork of several industry visionaries and experts. Some of the main actions required from are listed below.

- **Attract new members.** Historically, the TPC benchmark is represented by traditional systems and database vendors. While many of the existing member companies have Big Data practices, the TPC must attract new members companies in the Big Data management space including Hadoop, NoSQL and analytics. Based on a survey conducted at WBDB 2013, several companies have expressed their interest in joining industry standard consortia to develop industry standard benchmarks.
- **Focus on the fundamentals.** The TPC has a reputation of providing the industry with complete application level performance. There are five key aspects of a good benchmark articulated in the paper titled *The Art of Building a Good Benchmark* by Karl Huppler [14] as listed below.

    - Relevant – A reader of the result believes the benchmark reflects something important
    - Repeatable – There is confidence that the benchmark can be run a second time with the same result

- Fair – All systems and/or software being compared can participate equally
- Verifiable – There is confidence that the documented result is real
- Economical – The test sponsors can afford to run the benchmark.

Existing TPC benchmarks are designed meet these five aspects to some degree with clear strengths in some areas and accommodative of the others. As Big Data covers a broad range of technologies and products compromises on these aspects are required. It is possible that the TPC should consider a set of standards rather than one that can accommodate all technologies and products.

- **Build on existing workloads.** Historically the TPC has been developing standards from scratch. While this approach gives the full flexibility on the workload, metric and execution rules it is very expensive proposition in terms of resources from the TPC and member companies. Many members are not willing to bet on longer development cycles anymore because of the uncertainties on their return on resource investments. So the TPC should consider picking mature workloads in the academic and open source communities are add the TPC's core framework – metric, pricing and audit rules [16].
- **Collaboration with academia.** The TPC has made a major step in bringing the industry and academia together with its international conference series on performance evaluation and benchmarking (TPCTC) a concrete step towards identifying and fostering the development of the benchmarks of the future [7]. The conference series has been collocated with VLDB since 2009 which had attracted several papers. Benchmarking Big Data systems were one of the key topics identified at the TPCTC 2013 conference collocated with VLDB 2013 conference [9]. Collaboration with the academic community can speed up the development process.

4 Update from the TPC

The TPC continues to play a crucial role in providing the industry with relevant standards. The Workshop Series on Big Data Benchmarking (WBDB) has significantly influenced TPC's direction in looking at developing set of standards for benchmarking Big Data systems [19]. Also, Big Data was identified as one of the top areas for benchmark development at the most recent TPC Technology Conference on Performance Evaluation and Benchmarking [17]. In 2013 October the TPC formed a committee, represented by major server and software vendors, tasked to evaluate set of big data workloads and make directional recommendation to the TPC general council. Companies, research and government institutions who are interested in influencing the development of such benchmarks are encouraged to join the TPC [18].

5 Conclusion

Big Data is becoming an integral part of enterprise IT. Vendors, customer and researcher are challenged with tools and methodologies to compare software and hardware technologies dealing with Big Data. A set of Big Data benchmarks from the TPC is will not only provide means to compare performance, price-performance and energy efficiencies of both hardware and software systems but also instigate product developments and enhancements.

Acknowledgements. The author thanks the past and present members of the TPC.

References

1. Cisco Datacenter Blogs: Towards an Industry Standard for Benchmarking Big Data Workloads. http://blogs.cisco.com/datacenter/towards-an-industry-standard-for-benchmarking-big-data-workloads/. One More Step Closer Towards an Industry Standard for Benchmarking Big Data Workloads. http://blogs.cisco.com/datacenter/wbdb2012-in/
2. Nambiar, R., Poess, M.: A Review of System Benchmark Standards and a Look Ahead Towards an Industry Standard for Benchmarking Big Data Workloads. IGI Global, Hershey (2013)
3. Nambiar, R.O., Lanken, M., Wakou, N., Carman, F., Majdalany, M.: Transaction processing performance council (TPC): twenty years later – a look back, a look ahead. In: Nambiar, R., Poess, M. (eds.) TPCTC 2009. LNCS, vol. 5895, pp. 1–10. Springer, Heidelberg (2009)
4. IDC: Digital Universe Study (2010), IDC Worldwide Big Data Technology and Services 2012–2017 Forecast (2013)
5. Nambiar, R., Wakou, N., Carman, F., Majdalany, M.: Transaction processing performance council (TPC): state of the council 2010. In: Nambiar, R., Poess, M. (eds.) TPCTC 2010. LNCS, vol. 6417, pp. 1–9. Springer, Heidelberg (2011)
6. Nambiar, R., Poess, M., Masland, A., Taheri, H., Emmerton, M., Carman, F., Majdalany, M.: TPC benchmark roadmap 2012. In: Nambiar, R., Poess, M. (eds.) TPCTC 2012. LNCS, vol. 7755, pp. 1–20. Springer, Heidelberg (2013)
7. TPC Bylaws. http://www.tpc.org/information/about/documentation/Bylaws_v2.7.htm. TPC Policies. http://www.tpc.org/information/about/documentation/TPC_Policies_v6.0.htm
8. Nambiar, R., Poess, M. (eds.): TPCTC 2009. LNCS, vol. 5895, pp. 1–10. Springer, Heidelberg (2009)
9. Nambiar, R., Poess, M. (eds.): TPCTC 2012. LNCS, vol. 7755, pp. 1–20. Springer, Heidelberg (2013)
10. Nambiar, R., Poess, M.: Keeping the TPC relevant! PVLDB 6(11), 1186–1187 (2013)
11. Nambiar, R., Poess, M.: Transaction performance vs. Moore's law: a trend analysis. In: Nambiar, R., Poess, M. (eds.) TPCTC 2010. LNCS, vol. 6417, pp. 110–120. Springer, Heidelberg (2011)
12. Poess, M., Nambiar, R., Walrath, D.: Why you should run TPC-DS: a workload analysis. In: VLDB 2007, pp. 1138–1149 (2007)
13. Nambiar, R., Poess, M.: The making of TPC-DS. In: VLDB 2006, pp. 1049–1058
14. Huppler, K.: The art of building a good benchmark. In: Nambiar, R., Poess, M. (eds.) TPCTC 2009. LNCS, vol. 5895, pp. 18–30. Springer, Heidelberg (2009)
15. Gartner Survey (2013)
16. Stonebraker, M.: A new direction for TPC? In: Nambiar, R., Poess, M. (eds.) TPCTC 2009. LNCS, vol. 5895, pp. 11–17. Springer, Heidelberg (2009)

17. Nambiar, R., Poess, M. (eds.): TPCTC 2013. LNCS, vol. 8391, pp. 1–15. Springer, Heidelberg (2014)
18. Join the TPC. http://www.tpc.org/information/about/join.asp
19. Rabl, T., Posses, M., Baru, C., Jacobsen, H.-A. (eds.): WBDB 2012. LNCS, vol. 8163, pp. 1–10. Springer, Heidelberg (2014)

Author Index

Printed in the United States
By Bookmasters